Matrix Methods and
Fractional Calculus

Matrix Methods and Fractional Calculus

Arak M Mathai

McGill University, Canada

Hans J Haubold

*UN Office for Outer Space Affairs,
Vienna International Centre, Austria*

EW JERSEY • LONDON • SINGAPORE • BEIJING • SHANGHAI • HONG KONG • TAIPEI • CHENNAI • TOKYO

Published by

World Scientific Publishing Co. Pte. Ltd.

5 Toh Tuck Link, Singapore 596224

USA office: 27 Warren Street, Suite 401-402, Hackensack, NJ 07601

UK office: 57 Shelton Street, Covent Garden, London WC2H 9HE

Library of Congress Cataloging-in-Publication Data
Names: Mathai, A. M., author. | Haubold, H. J., author.
Title: Matrix methods and fractional calculus / by
  Arak M. Mathai (McGill University, Canada), Hans J. Haubold
  (UN Office for Outer Space Affairs, Vienna International Centre, Austria).
Description: New Jersey : World Scientific, 2017. | Includes bibliographical references.
Identifiers: LCCN 2017026674 | ISBN 9789813227521 (hc : alk. paper)
Subjects: LCSH: Matrices--Textbooks. | Fractional calculus--Textbooks.
Classification: LCC QA188 .M3855 2017 | DDC 515/.83--dc23
LC record available at https://lccn.loc.gov/2017026674

British Library Cataloguing-in-Publication Data
A catalogue record for this book is available from the British Library.

Desk Editors: V. Vishnu Mohan/Tan Rok Ting

Typeset by Stallion Press
Email: enquiries@stallionpress.com

Printed in Singapore

Preface

This book is an abridged version of the lectures given at the 2014 and 2015 SERB Schools at CMSS. For the 2014 and 2015 SERB Schools the topic was Matrix Methods and Fractional Calculus. Foreign lecturers included Professor Francesco Mainardi from Italy and Professor Serge B. Provost from Canada. Professor Rudolf Gorenflo from Germany and Professor Hans J. Haubold from Austria and the United Nations were supposed to come to give lectures but due to unexpected health problems they could not make it. Indian lecturers included Professors V. Daftardar-Gejji from Pune, M.A. Pathan from CMSS and Aligarh, R.B. Bapat from New Delhi and N. Mukunda from Bengaluru. The course Director is Professor A.M. Mathai (India/Canada) who is also the Director of CMSS. Professor N. Mukunda lectured on the applications of Hermitian positive definite matrices in quantum mechanics, light scattering and other areas and Professor R.B. Bapat lectured on matrix methods in graph theory. These two lecturers did not make available their lecture notes and hence they are not included in the current book. Professor A.M. Mathai lectured on mathematical and statistical preliminaries, matrix-variate statistical distributions, functions of matrix argument etc. Professor Mainardi lectured on the analysis aspects of fractional calculus and Professor Daftardar-Gejji lectured on the applications of fractional differential equations in control theory and engineering problems. Professor Pathan's lectures were on Lie Groups, Lie Algebra connected with special functions and Professor S.B. Provost from

v

the University of Western Ontario, Canada, lectured on some aspects of multivariate statistical analysis.

Chapters 1 and 2 provide the mathematical and statistical preliminaries, vector and matrix differential operators, their applications in quadratic and bilinear forms, maxima/minima problems, optimizations, Jacobians of matrix transformations and functions of matrix argument. Chapter 3 is on the theory of fractional integrals and fractional derivatives, Mittag-Leffler functions and their properties. Mittag-Leffler function is considered as the queen function in fractional calculus. Chapter 4 gives fractional differential equations, applications of fractional calculus in engineering and control theory problems, Adomian decomposition and iterative methods for the solutions of fractional differential equations. Chapter 5 is on the recent developments on matrix-variate fractional integrals and fractional derivatives or fractional calculus for functions of matrix argument. Chapter 6 is on Lie theory and special functions. Chapter 7 is on some aspects of multivariate statistical analysis, multivariate Gaussian and Wishart densities, their properties, tests of statistical hypotheses etc.

The material is useful for people who do research work in the areas of special functions, fractional calculus, applications of fractional calculus, and mathematical statistics, especially the multivariate and matrix-variate statistical distributions, tests of hypotheses, etc. Since the material is based on lecture notes, the material is also good for someone to get into these areas for their research work.

Dr H.J. Haubold was an integral part of these SERC Schools, one of the organizers and one of the foreign lecturers. Then A.M. Mathai and H.J. Haubold organized a national level conference on fractional calculus in 2012 at CMSS Pala Campus. Then the lecture notes were updated and brought out as Module 10 of CMSS in August 2014. A second printing of Module 10 took place in 2015.

The 2014 SERB Notes were brought out in the Publications Series of CMSS as Publication Number 44 of CMSS. Publications Series of CMSS consist of research level books and monographs. Since there was much overlap with the material of 2014 SERB School, no separate publication was brought out for the 2015 SERB

Notes. Research level Publication Number 44 of CMSS was developed with financial support from DST, Government of India, New Delhi, under Project Number SR/S4/MS:783/12. The authors would like to express their sincere gratitude to DST, Government of India, New Delhi, for the financial assistance.

A.M. Mathai

Hans J. Haubold

Peechi, Kerala, India

2nd April 2017

Acknowledgments

This is a modified version of CMSS (Centre for Mathematical and Statistical Sciences) 2014 and 2015 SERB School Notes. SERB Schools (Science and Engineering Research Board of the Department of Science and Technology, Government of India) are annual four weeks intensive all-India research level course on the topic. CMSS (formerly CMS = Centre for Mathematical Sciences) has been conducting these Schools from 1995 onward. In the earlier years the Schools used to be of six weeks duration but from 2005 onward the duration was cut down to four weeks. The first sequence of Schools was on Special Functions and Their Applications. Summarized version of these notes was brought out as the Springer, New York, publication *Special Functions for Applied Scientists* in 2008. The second sequence was on Functions of Matrix Argument and Their Applications. The third sequence was on Matrix Methods and Fractional Calculus. From 2006 onward, Fractional Calculus became an integral part of the SERC (Science and Engineering Research Council, later became research board or SERB) Schools at CMSS.

These CMSS SERB Notes are printed at CMSS Press and published by CMSS. Copies are made available to students free of cost and to researchers and others at production cost. For the preparation and printing of these publications, financial assistance was available from the Department of Science and Technology, Government of India (DST), New Delhi under project number SR/S4/MS:783/12.

Hence the authors would like to express their thanks and gratitude to DST, Government of India, for the financial assistance.

A.M. Mathai

Hans J. Haubold

Peechi, Kerala, India

2nd April 2017

Contents

List of Symbols

$\lvert(\cdot)\rvert$	absolute value/determinant of $(\cdot)$/Section 1.1, p. 1
$\text{tr}(X)$	trace of the matrix X/Section 1.1, p. 1
dX	wedge product of differentials/Section 1.1, p. 2
J	Jacobian/Section 1.1, p. 3
$X > O, X \geq O$	definiteness of matrices/Section 1.1, p. 6
$X < O, X \leq O$	definiteness of matrices/Section 1.1, p. 6
$\int_A^B f(X)dX$	integral over all X such that/Section 1.1, p. 6 $X > O, X - A > O, B - X > O$/Section 1.1, p. 6
X^*	conjugate transpose of X/Section 1.1, p. 6
$\frac{\partial}{\partial X}$	vector differential operator/Section 1.2, p. 7
X'	transpose the matrix X/Section 1.2, p. 7
$\text{diag}(a_{11}, \ldots, a_{pp})$	diagonal matrix/Section 1.4, p. 41
$\otimes$	Kronecker product/Section 2.3, p. 71
$\Gamma_p(\alpha)$	real matrix-variate gamma/Section 2.4, p. 74
$B_p(\alpha, \beta)$	real matrix-variate beta/Section 2.4, p. 76
(dX)	matrix of differentials dx_{ij}'s/Section 2.5, p. 86
$\tilde{\Gamma}_p(\alpha)$	complex matrix-variate gamma/Eq. (2.6.7), p. 97
I_{a+}^{α}	left-sided fractional integral/Eq. (3.2.3), p. 107
I_{b-}^{α}	right-sided fractional integral/Eq. (3.2.5), p. 108

D^n	integer-order derivative/Section 3.2, p. 109
D_{a+}^α	left-sided fractional derivative/Section 3.2, p. 110
D_{b-}^α	right-sided fractional derivative/Section 3.2, p. 110
$E_\alpha(z)$	Mittag-Leffler functions/Eq. (3.6.1), p. 129
$E_{\alpha,\beta}(z)$	Mittag-Leffler function/Eq. (3.6.5), p. 130
$\text{erf}(z)$	error function/Eq. (3.6.4), p. 129
$\Phi_{\lambda,\mu}(z)$	Wright function/Eq. (3.7.1), p. 143
$J_\nu^{(\lambda)}(z)$	Wright generalized Bessel function/Eq. (3.7.3), p. 145
$^c D^\alpha$	Caputo fractional derivative/Eq. (3.5.11), p. 124
$K_{2,u,\gamma}^{-\alpha}$	Kober integral, second kind/Eq. (5.3.2), p. 207
$K_{1,u,\gamma}^{-\alpha}$	Kober integral, first kind/Eq. (5.4.1), p. 209
$C_K(Z)$	zonal polynomial/Section 5.4, p. 211
$(a)_K$	generalized Pochhmmer symbol/Eq. (5.4.7), p. 211
$\tilde{D}_1^\alpha, \tilde{D}_2^\alpha$	fractional derivatives, matrix-variate/ Section 5.5, p. 214
e^A	matrix series/Definition 6.2.1, p. 220
$\text{GL}(2,C)$	general linear group/Example 6.5.3, p. 232
j^+, j^-	Lie group elements/Example 6.5.3, p. 232
$f_{x_1,...,x_k}(X)$	multivariate density/Section 7.2.1, p. 248
$F_{x_1,...,x_k}(X)$	multivariate distribution function/Eq. (7.2.1), p. 248
$E[g(X)]$	expected value/Section 7.2.1, p. 250
$N_p(\mu, V)$	multivariate normal density/Section 7.2.3, p. 253

Chapter 1

Vector/Matrix Derivatives and Optimization*

1.1. Introduction

The type of functions that we are going to deal with in the present series of lectures is real-valued scalar functions where the argument of the function can be a scalar variable, a vector variable or a matrix variable. To start with, we will consider the argument to be real, real scalar variable or real matrix variable in the sense that the elements of the matrix are real variables or real constants, not complex variables. Later we will consider the situations where the argument is a matrix defined over the complex domain.

The following standard notations will be used: Matrices will be denoted by capital letters, variable matrices by $X, Y, Z, \ldots$ and constant matrices by $A, B, C, \ldots$ scalar variables by $x, y, z, \ldots$ and scalar constants by $a, b, \ldots$. If $X = (x_{ij})$ is a $p \times p$ matrix then $\text{tr}(X) =$ trace of $X =$ sum of the leading diagonal elements $= x_{11} + x_{22} + \cdots + x_{pp} = \lambda_1 + \lambda_2 \cdots + \lambda_p$ where $\lambda_1, \lambda_2, \ldots, \lambda_p$ are the eigenvalues of X. The determinant of X will be denoted by $|X|$ or $\det(X)$. For example, if the determinant is $a + ib = \det(X)$ where, a, b real and $i = \sqrt{-1}$ then the conjugate is $a - ib$ and the absolute value is

$$|a + ib| = |a - ib| = [(a + ib)(a - ib)]^{\frac{1}{2}} = (a^2 + b^2)^{\frac{1}{2}} = |\det(X)|.$$

---

*This chapter is summarized from the lectures given by Professor Dr A.M. Mathai.

1

When $f(X)$ is a real-valued scalar function of the $m \times n$ matrix X, then $\int_X f(X)\mathrm{d}X$ will mean the integral over all $m \times n$ matrices. Here $\mathrm{d}X$ stands for the wedge product of differentials, that is,

$$\mathrm{d}X = \bigwedge_{i=1}^{m} \bigwedge_{j=1}^{n} \mathrm{d}x_{ij}, \qquad (1.1.1)$$

where $\wedge$ = wedge, $\mathrm{d}x \wedge \mathrm{d}y$ is the wedge product or skew symmetric product of differentials $\mathrm{d}x$ and $\mathrm{d}y$ and the skew symmetric product is defined as

$$\mathrm{d}x \wedge \mathrm{d}y = -\mathrm{d}y \wedge \mathrm{d}x \Rightarrow \mathrm{d}x \wedge \mathrm{d}x = 0, \quad \mathrm{d}y \wedge \mathrm{d}y = 0. \qquad (1.1.2)$$

Let us see the consequence of this definition of wedge product. Let $f_1(x_1, x_2)$ and $f_2(x_1, x_2)$ be two real-valued scalar functions of the real scalar variables x_1 and x_2. Let $y_1 = f_1(x_1, x_2)$ and $y_2 = f_2(x_1, x_2)$. Then from basic calculus, the differentials of y_1 and y_2 are given by

(i) $\quad \mathrm{d}y_1 = \dfrac{\partial f_1}{\partial x_1}\mathrm{d}x_1 + \dfrac{\partial f_1}{\partial x_2}\mathrm{d}x_2,$

(ii) $\quad \mathrm{d}y_2 = \dfrac{\partial f_2}{\partial x_1}\mathrm{d}x_1 + \dfrac{\partial f_2}{\partial x_2}\mathrm{d}x_2.$

Taking wedge product, we have

(iii) $\quad \mathrm{d}y_1 \wedge \mathrm{d}y_2 = \left[\dfrac{\partial f_1}{\partial x_1}\mathrm{d}x_1 + \dfrac{\partial f_1}{\partial x_2}\mathrm{d}x_2\right] \wedge \left[\dfrac{\partial f_2}{\partial x_1}\mathrm{d}x_1 + \dfrac{\partial f_2}{\partial x_2}\mathrm{d}x_2\right].$

Straight multiplication of the right side of (iii), keeping the order, gives

(iv) $\quad \mathrm{d}y_1 \wedge \mathrm{d}y_2 = \dfrac{\partial f_1}{\partial x_1}\dfrac{\partial f_2}{\partial x_1}\mathrm{d}x_1 \wedge \mathrm{d}x_1 + \dfrac{\partial f_1}{\partial x_1}\dfrac{\partial f_2}{\partial x_2}\mathrm{d}x_1 \wedge \mathrm{d}x_2$

$$+ \dfrac{\partial f_1}{\partial x_2}\dfrac{\partial f_2}{\partial x_1}\mathrm{d}x_2 \wedge \mathrm{d}x_1 + \dfrac{\partial f_1}{\partial x_2}\dfrac{\partial f_2}{\partial x_2}\mathrm{d}x_2 \wedge \mathrm{d}x_2$$

$$= 0 + \dfrac{\partial f_1}{\partial x_1}\dfrac{\partial f_2}{\partial x_2}\mathrm{d}x_1 \wedge \mathrm{d}x_2 + \dfrac{\partial f_1}{\partial x_2}\dfrac{\partial f_2}{\partial x_1}\mathrm{d}x_2 \wedge \mathrm{d}x_1 + 0$$

$$= \left[\frac{\partial f_1}{\partial x_1}\frac{\partial f_2}{\partial x_2} - \frac{\partial f_1}{\partial x_2}\frac{\partial f_2}{\partial x_1}\right] dx_1 \wedge dx_2 \text{ from (1.1.2)}$$

$$= \begin{vmatrix} \dfrac{\partial f_1}{\partial x_1} & \dfrac{\partial f_1}{\partial x_2} \\ \dfrac{\partial f_2}{\partial x_1} & \dfrac{\partial f_2}{\partial x_2} \end{vmatrix} dx_1 \wedge dx_2.$$

The coefficient is the determinant of the matrix $\left(\frac{\partial f_i}{\partial x_j}\right)$, where the (i,j)th element is the partial derivative of f_i with respect to x_j. From the structure of (iv) it is evident that in the general case when

$$y_1 = f_1(x_1, \ldots, x_k)$$
$$y_2 = f_2(x_1, \ldots, x_k)$$
$$\vdots$$
$$y_k = f_k(x_1, \ldots, x_k),$$

then the connection between the wedge product of differentials in $y_1, \ldots, y_k$ and $x_1, \ldots, x_k$, denoted by dY and dX respectively, is given by the following:

$$dy_1 \wedge \cdots \wedge dy_k = \begin{vmatrix} \dfrac{\partial f_1}{\partial x_1} & \cdots & \dfrac{\partial f_1}{\partial x_k} \\ \vdots & \cdots & \vdots \\ \dfrac{\partial f_k}{\partial x_1} & \cdots & \dfrac{\partial f_k}{\partial x_k} \end{vmatrix} dx_1 \wedge \cdots \wedge dx_k.$$

That is, observing that $dY = dY' = dy_1 \wedge \cdots \wedge dy_k$, $Y' = (y_1, \ldots, y_k) = $ transpose of Y, and $dX = dX' = dx_1 \wedge \cdots \wedge dx_k$, we have

$$dY = J dX, \quad J = \det\left(\frac{\partial f_i}{\partial x_j}\right) = \text{Jacobian}, \tag{1.1.3}$$

where J is called the Jacobian of the transformation of X going to Y.

Example 1.1.1. Let $x_1 = r\cos^2\theta, x_2 = r\sin^2\theta, 0 \leq r < \infty, 0 \leq \theta \leq \frac{\pi}{2}$. Let $dX = d(x_1, x_2) = dx_1 \wedge dx_2, d(r, \theta) = dr \wedge d\theta$. Then compute the Jacobian of this transformation.

Solution 1.1.1.

$$\frac{\partial x_1}{\partial r} = \cos^2\theta, \quad \frac{\partial x_1}{\partial \theta} = -2r\cos\theta\sin\theta,$$

$$\frac{\partial x_2}{\partial r} = \sin^2\theta, \quad \frac{\partial x_2}{\partial \theta} = 2r\cos\theta\sin\theta.$$

Then

$$dx_1 \wedge dx_2 = \begin{vmatrix} \dfrac{\partial x_1}{\partial r} & \dfrac{\partial x_1}{\partial \theta} \\ \dfrac{\partial x_2}{\partial r} & \dfrac{\partial x_2}{\partial \theta} \end{vmatrix} dr \wedge d\theta = \begin{vmatrix} \cos^2\theta & -2r\cos\theta\sin\theta \\ \sin^2\theta & 2r\cos\theta\sin\theta \end{vmatrix} dr \wedge d\theta$$

$$= \{\cos^2\theta[2r\cos\theta\sin\theta] + \sin^2\theta[2r\cos\theta\sin\theta]\}dr \wedge d\theta$$

$$= \{2r\cos\theta\sin\theta[\cos^2\theta + \sin^2\theta]\}dr \wedge d\theta$$

$$= 2r\cos\theta\sin\theta\, dr \wedge d\theta$$

$$= J\, d(r, \theta), \quad J = 2r\cos\theta\sin\theta.$$

Example 1.1.2. Let $y_1 = 2x_1 + x_2$ and $y_2 = x_1 - 2x_2$. Evaluate $dy_1 \wedge dy_2$ in terms of $dx_1 \wedge dx_2$.

Solution 1.1.2. We may either evaluate dy_1 and dy_2 separately and then take the wedge product or use the result in terms of determinant. Let us work out by using both the procedures: $dy_1 = 2dx_1 + dx_2$ and $dy_2 = dx_1 - dx_2$. Then

$$dy_1 \wedge dy_2 = [2dx_1 + dx_2] \wedge [dx_1 - 2dx_2]$$

$$= 2dx_1 \wedge dx_1 + 2(-2)dx_1 \wedge dx_2 + dx_2 \wedge dx_1 - 2dx_2 \wedge dx_2$$

$$= -4dx_1 \wedge dx_2 + dx_2 \wedge dx_1$$

$$= -4dx_1 \wedge dx_2 - dx_1 \wedge dx_2 = -5dx_1 \wedge dx_2.$$

Using determinant

$$dy_1 \wedge dy_2 = \left| \left(\frac{\partial y_i}{\partial x_j} \right) \right| dx_1 \wedge dx_2 = \begin{vmatrix} \dfrac{\partial y_1}{\partial x_1} & \dfrac{\partial y_1}{\partial x_2} \\ \dfrac{\partial y_2}{\partial x_1} & \dfrac{\partial y_2}{\partial x_2} \end{vmatrix} dx_1 \wedge dx_2$$

$$= \begin{vmatrix} 2 & 1 \\ 1 & -2 \end{vmatrix} dx_1 \wedge dx_2 = -5 dx_1 \wedge dx_2.$$

Example 1.1.3. Consider the elementary symmetric functions $y_1 = x_1 + x_2 + x_3$, $y_2 = x_1 x_2 + x_1 x_3 + x_2 x_3$, $y_3 = x_1 x_2 x_3$. Compute $dy_1 \wedge dy_2 \wedge dy_3$ in terms of $dx_1 \wedge dx_2 \wedge dx_3$.

Solution 1.1.3. $dy_1 = dx_1 + dx_2 + dx_3$, $dy_2 = x_1 dx_2 + x_1 dx_3 + x_2 dx_1 + x_3 dx_1 + x_2 dx_3 + x_3 dx_2$, $dy_3 = x_1 x_2 dx_3 + x_1 x_3 dx_2 + x_2 x_3 dx_1$.

$$dy_1 \wedge dy_2 = [dx_1 + dx_2 + dx_3] \wedge [x_1 dx_2 + x_2 dx_1 + \cdots + x_2 dx_3]$$

$$= x_1 dx_1 \wedge dx_2 + 0 + x_1 dx_1 \wedge dx_3 + 0$$

$$+ x_2 dx_1 \wedge dx_3 + x_3 dx_1 \wedge dx_2$$

$$+ 0 - x_2 dx_1 \wedge dx_2 + x_1 dx_2 \wedge dx_3$$

$$- x_3 dx_1 \wedge dx_2 + x_2 dx_1 \wedge dx_3 + 0$$

$$- x_1 dx_2 \wedge dx_3 - x_2 dx_1 \wedge dx_3 + 0$$

$$- x_3 dx_1 \wedge dx_3 + 0 - x_3 dx_2 \wedge dx_3.$$

Now, take

$$dy_1 \wedge dy_2 \wedge dy_3 = [dy_1 \wedge dy_2] \wedge dy_3$$

$$= [x_1^2 x_2 - x_1 x_2^2 - x_1^2 x_3 + x_1 x_3^2 + x_2^2 x_3 - x_2 x_3^2]$$

$$\times dx_1 \wedge dx_2 \wedge dx_3$$

$$= (x_1 - x_2)(x_1 - x_3)(x_2 - x_3) dx_1 \wedge dx_2 \wedge dx_3.$$

Now, by using determinant we have

$$\left| \left(\frac{\partial y_i}{\partial x_j} \right) \right| = \begin{vmatrix} 1 & 1 & 1 \\ x_2 + x_3 & x_1 + x_3 & x_1 + x_2 \\ x_2 x_3 & x_1 x_3 & x_1 x_2 \end{vmatrix}$$

$$= \begin{vmatrix} 1 & 1 & 1 \\ 0 & x_1 - x_2 & x_1 - x_3 \\ 0 & x_1 x_3 - x_2 x_3 & x_1 x_2 - x_2 x_3 \end{vmatrix}$$

$$= \begin{vmatrix} 1 & 0 & 0 \\ 0 & x_1 - x_2 & x_1 - x_3 \\ 0 & x_3(x_1 - x_2) & x_2(x_1 - x_3) \end{vmatrix}$$

$$= (x_1 - x_2)(x_1 - x_3) \begin{vmatrix} 1 & 0 & 0 \\ 0 & 1 & 1 \\ 0 & x_3 & x_2 \end{vmatrix}$$

$$= (x_1 - x_2)(x_1 - x_3)(x_2 - x_3).$$

Let us continue with our notations. When X is $p \times p$ real and positive definite then we will denote it by $X > O$. Observe that definiteness is defined only for symmetric matrices, $X = X'$, when real and hermitian matrices, $X = X^*$, when in the complex domain where an $*$ indicates complex conjugate transpose. Similarly,

$X > O$ (X is positive definite),

$X \geq O$ (X is positive semidefinite),

$X < O$ (X is negative definite),

$X \leq O$ (X is negative semidefinite).

Then $O < A < X < B$ will mean A, X, B are positive definite, $X - A$ is positive definite, $B - X$ is positive definite or $X - A > O, B - X > O, X > O, A > O, B > O$. Moreover,

$$\int_A^B f(X) \mathrm{d}X = \int_{A<X<B} f(X) \mathrm{d}X = \int_{O<A<X<B} f(X) \mathrm{d}X \quad (1.1.4)$$

will mean the integral of the real-valued scalar function $f(X)$ of the positive definite matrix X, over all X, such that $X > O, X - A > O, A > O, B - X > O, B > O$, where $\mathrm{d}X$ is the wedge product of the differentials in X. For example, $\int_{O<X<I} f(X) \mathrm{d}X$ means the integral over all X such that $X > O, I - X > O$ or all eigenvalues of X are in the open interval $(0, 1)$.

Exercise 1.1

1.1.1. Evaluate $dy_1 \wedge dy_2$ in terms of $dx_1 \wedge dx_2$ from first principles as well as by using determinant when $y_1 = x_1^2 + x_2^2, y_2 = x_1^2 + 5x_1 x_2 + x_2^2$.

1.1.2. Let $y_1 = x_1 + x_2$ and $y_2 = x_1^2 + 2x_2^2 + x_1 x_2$. Evaluate $dy_1 \wedge dy_2$ in terms of $dx_1 \wedge dx_2$ by evaluating dy_1 and dy_2 and then taking $dy_1 \wedge dy_2$ as well as by using determinant.

1.1.3. Let $y_1 = x_1 + x_2 + x_3, y_2 = 2x_1 - 3x_2 + x_3, y_3 = x_1 - x_2 + x_3$. Evaluate $dy_1 \wedge dy_2 \wedge dy_3$ in terms of $dx_1 \wedge dx_2 \wedge dx_3$ from first principles as well as by using determinant.

1.1.4. Let $x_1 = r \cos\theta_1 \cos\theta_2, x_2 = r \cos\theta_1 \sin\theta_2, x_3 = r \sin\theta_1$. Evaluate $dx_1 \wedge dx_2 \wedge dx_3$ in terms of $dr \wedge d\theta_1 \wedge d\theta_2$ from first principles as well as by using determinant.

1.1.5. Let $y_1 = x_1 + x_2 + x_3, y_2 = x_1^2 + x_2^2 + x_3^2, y_3 = x_1^3 + x_2^3 + x_3^3$. Evaluate $dy_1 \wedge dy_2 \wedge dy_3$ in terms of $dx_1 \wedge dx_2 \wedge dx_3$.

1.2. A Vector Differential Operator

Let X be a $p \times 1$ vector of real scalar variables, then let

$$\frac{\partial}{\partial X} = \begin{bmatrix} \dfrac{\partial}{\partial x_1} \\ \vdots \\ \dfrac{\partial}{\partial x_p} \end{bmatrix}, \quad \frac{\partial}{\partial X'} = \begin{bmatrix} \dfrac{\partial}{\partial x_1}, \dots, \dfrac{\partial}{\partial x_p} \end{bmatrix} \qquad (1.2.1)$$

or $\frac{\partial}{\partial X}$ is the column vector of partial differential operators and $\frac{\partial}{\partial X'}$ is its transpose operator. Then

$$\frac{\partial}{\partial X} \frac{\partial}{\partial X'} = \begin{bmatrix} \dfrac{\partial^2}{\partial x_1^2} & \dfrac{\partial^2}{\partial x_1 \partial x_2} & \cdots & \dfrac{\partial^2}{\partial x_1 \partial x_p} \\ \vdots & \vdots & \vdots & \vdots \\ \dfrac{\partial^2}{\partial x_p \partial x_1} & \dfrac{\partial^2}{\partial x_p \partial x_2} & \cdots & \dfrac{\partial^2}{\partial x_p^2} \end{bmatrix}. \qquad (1.2.2)$$

Let us see the effect of the operators in (1.2.1) and (1.2.2) operating on a scalar function.

1.2.1. Linear forms

A linear form is a homogeneous function of degree 1 or a linear function where every term is of degree one each, and a linear expression is where the maximum degree is one. For example,

$$u_1 = 2x_1 - 5x_2 + x_3 \quad \text{(linear form)},$$

$$u_2 = x_1 + 3x_2 - x_3 + x_4 + 7 \quad \text{(linear expression)},$$

$$u_3 = x_1 + \cdots + x_k \quad \text{(linear form)},$$

$$u_4 = a_1 x_1 + \cdots + a_k x_k \quad \text{(linear form, } a_1, \ldots, a_k \text{ constants)}.$$

A general linear form such as u_4 can be written as

$$u = a'X = X'a, \quad a' = (a_1, \ldots, a_k), \quad X' = (x_1, \ldots, x_k).$$

The differential operator $\frac{\partial}{\partial X}$ operating on u yields the following:

$$\frac{\partial u}{\partial X} = \begin{bmatrix} \dfrac{\partial u}{\partial x_1} \\ \vdots \\ \dfrac{\partial u}{\partial x_p} \end{bmatrix} = \begin{bmatrix} a_1 \\ \vdots \\ a_p \end{bmatrix} = a. \qquad (1.2.3)$$

Thus, we have the following result.

Result 1.2.1.

$$u = a'X = X'a \Rightarrow \frac{\partial u}{\partial X} = a.$$

1.2.2. Quadratic forms

A function of scalar variables $x_1, \ldots, x_p$ where every term is of degree two each is a quadratic form. If the maximum degree is 2 then the function is a quadratic expression which may consist of a quadratic term, a linear term and a constant also. For example,

$$q_1 = 2x_1^2 + 3x_2^2 - 5x_3^2 \quad \text{(quadratic form in } x_1, x_2),$$

$$q_2 = x_1^2 + 5x_2^2 - 7x_1x_2 + 3x_1 - x_2 + 5$$

$$\text{(quadratic expression in } x_1, x_2),$$

$$q_3 = x_1^2 + \cdots + x_p^2 \quad \text{(quadratic form)},$$

$$q_4 = a_{11}x_1^2 + \cdots + a_{pp}x_p^2 + \sum_{i \neq j} a_{ij}x_ix_j \quad \text{(quadratic form)},$$

where a_{ij}'s are constants, $x_1, \ldots, x_p$ are scalar variables. Note that q_4 can also be written by using vector-matrix notation. Moreover,

$$Q = X'AX, X = \begin{bmatrix} x_1 \\ \vdots \\ x_p \end{bmatrix}, \quad A = (a_{ij}) = \begin{bmatrix} a_{11} & \cdots & a_{1p} \\ \vdots & \ldots & \vdots \\ a_{p1} & \cdots & a_{pp} \end{bmatrix}.$$

For example

$$3x_1^2 + 5x_2^2 + x_3^2 - 2x_1x_2 + 4x_2x_3$$

$$= [x_1, x_2, x_3] \begin{bmatrix} 3 & -2 & 0 \\ 0 & 5 & 4 \\ 0 & 0 & 1 \end{bmatrix} \begin{bmatrix} x_1 \\ x_2 \\ x_3 \end{bmatrix}$$

$$= [x_1, x_2, x_3] \begin{bmatrix} 3 & -1 & 0 \\ -1 & 5 & 2 \\ 0 & 2 & 1 \end{bmatrix} \begin{bmatrix} x_1 \\ x_2 \\ x_3 \end{bmatrix}.$$

In the last form the matrix A is symmetric, $A = A'$. Whatever be A, when it appears in a quadratic form, we can always rewrite it as a symmetric matrix because

$$X'AX = (X'AX)' \text{ since } X'AX \text{ is a } 1 \times 1 \text{ matrix}$$

$$= X'A'X \Rightarrow X'AX = X'BX, B = \tfrac{1}{2}(A + A') = B'.$$

Hence, without loss of generality we may assume A to be symmetric when A is the matrix in a quadratic form. Let us see what happens if we operate with our operator $\frac{\partial}{\partial X}$. That is,

$$\frac{\partial Q}{\partial X} = \frac{\partial}{\partial X}[a_{11}x_1^2 + a_{12}x_1x_2 + \cdots + a_{1p}x_1x_p$$
$$+ a_{21}x_2x_1 + a_{22}x_2^2 + \cdots + a_{2p}x_2x_p$$
$$+ \cdots + a_{p1}x_px_1 + a_{p2}x_px_2 + \cdots + a_{pp}x_p^2].$$

Only the first row and first column in the above format contain x_1. Hence the partial derivative with respect to x_1 yields

$$\frac{\partial Q}{\partial x_1} = (a_{11}, a_{12}, \ldots, a_{1p})X + (a_{11}, a_{21}, \ldots, a_{p1})X.$$

Similarly the second row and second column contain x_2. Hence $\frac{\partial Q}{\partial x_2}$ yields

$$\frac{\partial Q}{\partial x_2} = (a_{21}, a_{22}, \ldots, a_{2p})X + (a_{12}, a_{22}, \ldots, a_{p2})X$$

and so on. Putting them in a column yields

$$\frac{\partial Q}{\partial X} = AX + A'X = \begin{cases} (A + A')X, \\ 2AX \quad \text{if } A = A'. \end{cases}$$

Therefore, we have the following result.

Result 1.2.2.

$$Q = X'AX, X' = (x_1, \ldots, x_p), \quad A = (a_{ij}) \Rightarrow \frac{\partial Q}{\partial X}$$

$$= \begin{cases} (A + A')X, \\ 2AX \quad \text{if } A = A'. \end{cases} \tag{1.2.4}$$

Example 1.2.1. Evaluate $\frac{\partial Q}{\partial X}$ where $Q = 2x_1^2 + x_2^2 - x_3^2 - 2x_1x_2 - 4x_1x_3$.

Solution 1.2.1. For convenience let us write Q as

$$Q = (x_1, x_2, x_3) \begin{bmatrix} 2 & -1 & -2 \\ -1 & 1 & 0 \\ -2 & 0 & -1 \end{bmatrix} \begin{bmatrix} x_1 \\ x_2 \\ x_3 \end{bmatrix} = X'AX, \quad A = A'.$$

Then

$$\frac{\partial Q}{\partial X} = \begin{bmatrix} \dfrac{\partial Q}{\partial x_1} \\[2mm] \dfrac{\partial Q}{\partial x_2} \\[2mm] \dfrac{\partial Q}{\partial x_3} \end{bmatrix} = \begin{bmatrix} 4x_1 & -2x_2 & -4x_3 \\ 2x_2 & -2x_1 & \\ -2x_3 & -4x_1 & \end{bmatrix}$$

$$= \begin{bmatrix} 4 & -2 & -4 \\ -2 & 2 & 0 \\ -4 & 0 & -2 \end{bmatrix} \begin{bmatrix} x_1 \\ x_2 \\ x_3 \end{bmatrix}$$

$$= 2 \begin{bmatrix} 2 & -1 & -2 \\ -1 & 1 & 0 \\ -2 & 0 & -1 \end{bmatrix} \begin{bmatrix} x_1 \\ x_2 \\ x_3 \end{bmatrix} = 2AX.$$

1.2.3. Quadratic form, hermitian form and definiteness

Optimization of quadratic form is often looked into in connection with maximization or minimization problems arising from various multivariate situations. The procedure eventually ends up with checking the definiteness of matrices or definiteness of the corresponding quadratic forms. Hence we will consider the concept of definiteness of matrices first.

1.2.3.1. *Definiteness of matrices and quadratic forms*

Definiteness is defined only for symmetric matrices $A = A'$ when the elements of A are real or hermitian matrices $A = A^*$ when the elements are in the complex domain, where A^* means the conjugate transpose of A. If $A = (a_{ij})$ then when $A = A'$ we have $a_{ij} = a_{ji}$ for all i and j, or the ith row jth column element is equal to the jth row ith column element. When $A = A^*$ we have $a_{ij} = \bar{a}_{ji}$ where a bar indicates the conjugate. For example,

$$A = \begin{bmatrix} 2 & 1 & -1 \\ 1 & 0 & 5 \\ -1 & 5 & -2 \end{bmatrix} = A',$$

where $a_{11} = 2, a_{12} = 1 = a_{21}, a_{13} = -1 = a_{31}, a_{23} = 5 = a_{32}, a_{22} = 0, a_{33} = -2$, and

$$B = \begin{bmatrix} 3 & 1+i & 2-3i \\ 1-i & 4 & i \\ 2+3i & -i & -2 \end{bmatrix} = B^*,$$

where $b_{11} = 3 = \bar{b}_{11}, b_{12} = 1 + i = \bar{b}_{21}, b_{13} = 2 - 3i = \bar{b}_{31}, b_{22} = 4 = \bar{b}_{22}, b_{23} = i = \bar{b}_{32}, b_{33} = -2 = \bar{b}_{33}$. Note that when B is hermitian the diagonal elements will be real because $b_{jj} = \bar{b}_{jj}$ which means that the imaginary part is zero. The standard notations used for definiteness are the following: $A = A'$ or $A = A^*$ and in the following O is the capital letter o and not zero:

$$A > O \quad \text{(positive definite)},$$

$$A \geq O \quad \text{(positive semidefinite)},$$

$$A < O \quad \text{(negative definite)},$$

$$A \leq O \quad \text{(negative semidefinite)},$$

and the matrices which do not belong to any of the above categories are called indefinite matrices. Let $A = (a_{ij})$ be a $p \times p$ matrix and let X be a $p \times 1$ vector where $X' = (x_1, \ldots, x_p)$. Let a_{ij}'s and x_{ij}'s be real. Consider the quadratic form $q = X'AX, A = A'$. If $q > 0$ for all non-null X then A is positive definite and the quadratic form $X'AX$ is called positive definite. Corresponding definitions for other cases are as follows:

$q = X'AX > 0$ for all $X \neq O$ (A is positive definite and $X'AX$ is positive definite),

$q = X'AX \geq 0$ for all $X \neq O$ (A and $X'AX$ are positive semi-definite),

$q = X'AX < 0$ for all $X \neq O$ (A and $X'AX$ are negative definite),

$q = X'AX \leq 0$ for all $X \neq O$ (A and $X'AX$ are negative semi-definite).

$$(1.2.5)$$

If $q > 0$ for some X and if $q < 0$ for some other X then A and q do not belong to any of the above categories and then A and q are called indefinite. For example, if $A = \begin{bmatrix} 3 & 0 \\ 0 & -1 \end{bmatrix}$ and if $X = \begin{bmatrix} 1 \\ 0 \end{bmatrix}$ then $X'AX = 3 > 0$. If $X = \begin{bmatrix} 0 \\ 1 \end{bmatrix}$ then $X'AX = -1 < 0$. Hence A here is indefinite. For example,

$$A_1 = \begin{bmatrix} 3 & 2 \\ 2 & 2 \end{bmatrix} > O, \quad A_2 = \begin{bmatrix} 2 & 2 \\ 2 & 2 \end{bmatrix} \geq O,$$

$$A_3 = \begin{bmatrix} -2 & 1 \\ 1 & -3 \end{bmatrix} < O, \quad A_4 = \begin{bmatrix} -2 & \sqrt{6} \\ \sqrt{6} & -3 \end{bmatrix} \leq O, \quad A_5 = \begin{bmatrix} 3 & 1 \\ 1 & -2 \end{bmatrix}$$

indefinite. Observe that if $A > O$ then all its diagonal elements must be positive. The converses need not be true. Similarly, if $A < O$ then all its diagonal elements must be negative. If some diagonal elements of A are positive and some diagonal elements of A are negative, then A has to be indefinite. These are evident by taking the vectors

$$X_1 = \begin{bmatrix} 1 \\ 0 \\ \vdots \\ 0 \end{bmatrix}, \quad X_2 = \begin{bmatrix} 0 \\ 1 \\ \vdots \\ 0 \end{bmatrix}, \ldots, X_p = \begin{bmatrix} 0 \\ 0 \\ \vdots \\ 1 \end{bmatrix}$$

and considering the quadratic form $X_j'AX_j = a_{jj}, j = 1, \ldots, p$.

How can we check whether a given quadratic form or a symmetric matrix is positive definite or not, or in general, how can we check the definiteness of a quadratic form? When $A = A'$ we know that there exists an orthonormal matrix $P, PP' = I, P'P = I$ such that $P'AP = \text{diag}(\lambda_1, \ldots, \lambda_p) = D = $ diagonal matrix with the diagonal elements being the eigenvalues of A. Also we know that when $A = A'$

all eigenvalues of A are real. In this case, $A = PDP'$ or the quadratic form

$$q = X'AX = X'PDP'X = Y'DY$$
$$= \lambda_1 y_1^2 + \cdots + \lambda_p y_p^2, \quad Y = P'X. \tag{1.2.6}$$

This means $X = PY$. Suppose that we would like to have $Y' = (1, 0, \ldots, 0)$ then choose X as the first column of P so that $Y' = (1, 0, \ldots, 0)$. In this case $q = X'AX = \lambda_1$. Hence $\lambda_1 > 0$ or $\lambda_j > 0, j = 1, \ldots, p$ if A is positive definite. We have the following rule.

Rule 1.2.1. Let $A = A'$ and all elements of A be real. Then $X'AX > 0$ for all nonnull X if and only if $\lambda_j > 0, j = 1, \ldots, p$ or if $A = A'$ then

$A > O$ (if and only if all eigenvalues of A are positive),

$A \leq O$ (if and only if the eigenvalues of A are positive or zero),

$$\tag{1.2.7}$$

$A < O$ (if and only if all eigenvalues of A are negative),

$A \leq O$ (eigenvalues of A are either negative or zero).

Note that A is indefinite if some eigenvalues of A are positive and some others are negative, others may be zeros. If $A = A'$ and if $\lambda_j > 0, j = 1, \ldots, p$ then retracing the steps from (1.2.6) and back we can see that $A'AX > 0$ for all non-null X. Similarly, we can show the converse in all other cases. Note that a triangular matrix with all diagonal elements positive is not a positive definite matrix. We can construct triangular matrices with all diagonal elements positive and at the same time we can construct an $X \neq O$ for which $X'AX$ is not positive. Hence $A = A'$ must be an essential condition to talk about definiteness.

Note that if $\lambda_j > 0, j = 1, \ldots, p$ then products taken two at a time, three at a time, $\ldots$, p at a time or product of all λ_j's is positive. Similarly, if $\lambda_j < 0, j = 1, \ldots, p$ then λ_j's taken one at a time is negative, products taken two at a time is positive, products taken three at a time is negative and so on, or product of odd number of times is negative and even number of times is positive. From these observations we can have the following rule.

Rule 1.2.2. Consider the leading minors of $A = A'$. The jth leading minor is the determinant of the jth leading submatrix or the submatrix obtained by deleting all rows and columns from $j + 1$ onward. For example,

$$a_{11} = \text{the first leading minor}$$

$$\begin{vmatrix} a_{11} & a_{12} \\ a_{12} & a_{22} \end{vmatrix} = \text{the second minor of } A$$

$$\vdots$$

$$|A| = \text{the } p\text{th minor of } A.$$

If $A = A'$ then $A > O$ if and only if all the leading minors are positive; $A \geq O$ if and only if the leading minors are positive or zero; $A < O$ if and only if the leading minors are negative, positive, etc. (odd order minors are negative, even order minors are positive); $A \leq O$ if and only if the leading minors are alternately negative and positive or zero; A is indefinite then the above rules are violated.

1.2.3.2. *Definiteness of hermitian forms*

If $A = A^*$, where A^* is the conjugate transpose of A, then A is hermitian. It is not difficult to show that all eigenvalues of A are real when $A = A^*$. Let λ be an eigenvalue of A and let X be a corresponding eigenvector which is normalized so that $X^*X = 1$. Then

$$AX = \lambda X \Rightarrow (AX)^* = (\lambda X)^* \Rightarrow X^*A^* = \bar{\lambda}X^*, \tag{a}$$

where $\bar{\lambda}$ is the conjugate of λ. Then, since $A = A^*$,

$$X^*A = \bar{\lambda}X^*. \tag{b}$$

Premultiply (a) by X^* and postmultiply (b) by X to get $X^*AX = \lambda X^*X = \lambda$ and $X^*AX = \bar{\lambda}X^*X = \bar{\lambda}$ or $\lambda = \bar{\lambda}$, which means that λ is real. This proof also holds for real symmetric case of A.

When A is hermitian there exists a unitary matrix $Q, QQ^* = I, Q^*Q = I$ such that $Q^*AQ = \text{diag}(\lambda_1, \ldots, \lambda_p) = D$ or $A = QDQ^*$. Therefore a hermitian form X^*AX can be written equivalently as

$$X^*AX = \lambda_1|y_1|^2 + \cdots + \lambda_p|y_p|^2, \quad Y = Q^*X \qquad (1.2.8)$$

and $|y_j|$ is the absolute value of $y_j, j = 1, \ldots, p, Y' = (y_1, \ldots, y_p)$.

Definiteness of hermitian forms is also defined parallel to that of a quadratic form and the notation remains the same:

$A > O$ or $X^*AX > 0$ for all $X \neq O$ (A is positive definite),

$A \geq O$ or $X^*AX \geq 0$ for all $X \neq O$ (A is positive semidefinite),

$A < O$ or $X^*AX < 0$ for all $X \neq O$ (A is negative definite),

$A \leq O$ or $X^*AX \leq 0$ for all $X \neq O$ (A is negative semidefinite).

If A does not belong to any of the above categories then A is indefinite, that is, for some non-null X if we have $X^*AX > 0$ and for some other $X \neq O$ if we have $X^*AX < 0$ then A and the corresponding hermitian form are indefinite. For checking definiteness, Rules 1.2.1 and 1.2.2 can also be applied here when $A = A^*$.

Exercise 1.2

1.2.1. Let $u_1 = 2x_1 - x_2 + x_3 - 2x_4, u_2 = 3x_1 + 2x_2 - x_3 - x_4, u_3 = x_1 - 5x_2 + x_3 - 2x_4$. Let $\frac{\partial}{\partial X}$ be the vector differential operator. Compute $\frac{\partial u_1}{\partial X}, \frac{\partial u_2}{\partial X}, \frac{\partial u_3}{\partial X}$.

1.2.2. Write the following quadratic forms in the form $X'AX$ where (i) A is symmetric, (ii) A is not symmetric: (a) $u_1 = x_1^2 + x_2^2 + x_3^2 + 2x_1x_2 - x_1x_3$, (b) $u_2 = 5x_1^2 + x_2^2 + 2x_3^2 - 2x_1x_3 + x_2x_3$, (c) $u_3 = 2x_1^2 + x_2^2 + 3x_3^2 - 2x_1x_2 + x_2x_3$.

1.2.3. Compute $\frac{\partial u_j}{\partial X}, j = 1, 2, 3$ in each case of A symmetric and A not symmetric in Exercise 1.2.2.

1.2.4. Write the following expression in the form $X'AX + X'b + c$ with $A = A'$: $u = 2x_1^2 + x_2^2 + 3x_3^2 - 2x_1x_2 + 5x_2 - x_1 + x_3 + 5$. Compute $\frac{\partial u}{\partial X}$.

1.2.5. Let $u = 2x_1^2 + 3x_2^2 - 2x_1x_2 + 5x_3^2 - 2x_1 + 2x_2 + x_3 + 7$. Convert u to the form $u = (X - \mu)'A(X - \mu) + c$ for some $A = A', \mu, c$. Then compute (i): $\frac{\partial u}{\partial(X-\mu)}$, (ii): $\frac{\partial u}{\partial X}$ and (iii): solve the equation $\frac{\partial U}{\partial X} = O$ (null vector).

1.2.6. Construct (1): a 2×2, (2): a 3×3 matrix of a quadratic form which is (a) positive definite, (b) positive semidefinite, (c) negative definite, (d) negative semidefinite, (e) indefinite. Matrix of the quadratic form should not be diagonal.

1.2.7. Repeat Exercise 1.2.6 with quadratic form replaced by hermitian form.

1.2.8. Check for maxima/minima of the following functions:

(a) $u = 2x_1^2 + 3x_2^2 + 5x_3^2 - 2x_1x_2$,
(b) $u_1 = u + 4x_1 - 2x_2 + 5x_3 - 7$,
(c) $u_2 = u$ subject to $2x_1 + 3x_2 + 5x_3 = 1$,
(d) $u_3 = u$ subject to $x_1^2 + x_2^2 + x_3^2 = 1$,
(e) $u_4 = u$ subject to $x_1^2 + 2x_2^2 + x_3^2 + x_1x_2 = 1$.

1.2.9. Optimize $2x_1 + 5x_2 - x_3$ subject to $x_1^2 + x_2^2 + x_3^2 = 1$.

1.2.10. Optimize $2x_1 + 5x_2 - x_3$ subject to $4x_1^2 + 3x_2^2 - 2x_1x_2 + 2x_3^2 + 2x_2x_3 = 1$.

1.3. Expansion of a Multivariate Function

Consider a real-valued scalar function $f(x_1, \ldots, x_k)$ of the real scalar variables $x_1, \ldots, x_k$. Suppose that we wish to expand $f(x_1, \ldots, x_k)$ near a point $a' = (a_1, \ldots, a_k)$. A small neighborhood of a may be denoted by $a + \delta$.

$$a + \delta = \begin{bmatrix} a_1 + \delta_1 \\ a_2 + \delta_2 \\ \vdots \\ a_k + \delta_k \end{bmatrix} = \begin{bmatrix} a_1 \\ a_2 \\ \vdots \\ a_k \end{bmatrix} + \begin{bmatrix} \delta_1 \\ \delta_2 \\ \vdots \\ \delta_k \end{bmatrix}, \quad \delta_j \to 0, j = 1, \ldots, k,$$

$$a = \begin{bmatrix} a_1 \\ a_2 \\ \vdots \\ a_k \end{bmatrix}, \quad \delta = \begin{bmatrix} \delta_1 \\ \delta_2 \\ \vdots \\ \delta_k \end{bmatrix}, \quad X = \begin{bmatrix} x_1 \\ x_2 \\ \vdots \\ x_k \end{bmatrix}.$$

For convenience, let us denote $f(a) = f(a_1, \ldots, a_k)$ and $f(a + \delta) = f(a_1 + \delta_1, \ldots, a_k + \delta_k)$. Then a power series expansion will be of the following form:

$$\begin{aligned} f(a + \delta) = b_{00\cdots 0} &+ [\delta_1 b_{10\cdots 0} + \cdots + \delta_k b_{0\cdots 1}] \\ &+ [\delta_1^2 b_{20\cdots 0} + \cdots + \delta_k^2 b_{0\ldots 02}, \\ &+ \delta_1 \delta_2 b_{110\cdots 0} + \cdots + \delta_{k-1} \delta_k b_{0\cdots 011}] + \cdots, \end{aligned}$$

where $b_{\ldots}$ are coefficients to be determined. We may also write

$$\begin{aligned} f(X) = f(x_1, \ldots, x_k) = b_{0\cdots 0} &+ [(x_1 - a_1) b_{10\cdots 0} \\ &+ \cdots + (x_k - a_k) b_{0\cdots 01}] + \cdots \end{aligned}$$

replacing $\delta_j = x_j - a_j, j = 1, \ldots, k$. What are these coefficients? Putting $\delta_1 = 0, \ldots, \delta_k = 0$ or $\delta = O$ on the right we have $b_{0\cdots 0} = f(a)$. Note that $\frac{\partial f}{\partial \delta_j}|_{\delta = O}, j = 1, \ldots, k$ yields $b_{10\cdots 0}, \ldots, b_{0\cdots 0j}$, etc. In order to construct the expansion as well as to evaluate the coefficients we will adopt the following procedure. Let $\delta' = (\delta_1, \ldots, \delta_k), D' = \frac{\partial}{\partial X'} = (\frac{\partial}{\partial x_1}, \ldots, \frac{\partial}{\partial x_k})$ and consider the dot product $\delta \cdot D$ or $\delta \cdot D = \delta' \frac{\partial}{\partial X}$ where the column vector operator $\frac{\partial}{\partial X}$ was already defined in Section 1.2. That is,

$$\delta = \begin{bmatrix} \delta_1 \\ \vdots \\ \delta_k \end{bmatrix}, \quad D = \frac{\partial}{\partial X} = \begin{bmatrix} \dfrac{\partial}{\partial x_1} \\ \vdots \\ \dfrac{\partial}{\partial x_k} \end{bmatrix},$$

$$\delta \cdot D = \delta' \frac{\partial}{\partial X} = \delta_1 \frac{\partial}{\partial x_1} + \cdots + \delta_k \frac{\partial}{\partial x_k}.$$

Let $e^{\delta' \frac{\partial}{\partial X}}$ denote an operator operating on f and then f is evaluated at $X = a$. But

$$e^{\delta' \frac{\partial}{\partial X}} = 1 + \frac{(\delta' \frac{\partial}{\partial X})}{1!} + \frac{(\delta' \frac{\partial}{\partial X})^2}{2!} + \cdots,$$

where, for example, $(\delta' \frac{\partial}{\partial X})^2 = \sum_{j=1}^{k} \delta_j^2 \frac{\partial^2}{\partial x_j^2} + 2 \sum_{i>j} \delta_i \delta_j \frac{\partial^2}{\partial x_i \partial x_j}$. Then $e^{\delta' \frac{\partial}{\partial X}}$ operating on f and then evaluating at $X = a$ has the following form:

$$f(a + \delta) = f(a_1 + \delta_1, \ldots, a_k + \delta_k)$$

$$= f(a) + \frac{1}{1!} \left[\sum_{j=1}^{k} \delta_j \frac{\partial f(a)}{\partial x_j} \right]$$

$$+ \frac{1}{2!} \left[\sum_{j=1}^{k} \delta_j^2 \frac{\partial^2 f(a)}{\partial x_j^2} + 2 \sum_{i<j} \delta_i \delta_j \frac{\partial^2 f(a)}{\partial x_i \partial x_j} \right] + \frac{1}{3!}[\ldots] + \cdots,$$

i.e.,

$$f(a + \delta) = e^{\delta' \frac{\partial}{\partial X}} f|_{X=a}. \qquad (1.3.1)$$

1.3.1. Maxima/minima of functions of many variables

Note that (1.3.1) can also be written as follows, where $\frac{\partial^r f(a)}{\partial x_j^r} = \frac{\partial^r f(X)}{\partial x_j^r}\Big|_{X=a}$ or evaluated at $X = a$:

$$f(a + \delta) - f(a) = \frac{1}{1!} \left[\sum_{j=1}^{k} \delta_j \frac{\partial f(a)}{\partial x_j} \right] + \frac{1}{2!}[\ldots] + \cdots.$$

Suppose $X = a$ is a critical point where f has either a local maximum or a local minimum or a saddle point. At a critical point, all first-order derivatives will vanish. Then

$$f(a + \delta) - f(a) = \frac{1}{2!} \left[\sum_{j=1}^{k} \delta_j^2 \frac{\partial^2 f(a)}{\partial x_j^2} + 2 \sum_{i<j} \delta_i \delta_j \frac{\partial^2 f(a)}{\partial x_i \partial x_j} \right]$$

$$+ \frac{1}{3!}[\ldots] + \cdots.$$

When $\delta \to O$ then the leading term on the right will determine the sign of the right side. The leading term on the right is the second-order term, namely,

$$\frac{1}{2!}\left[\sum_{j=1}^{k}\delta_j^2\frac{\partial^2 f(a)}{\partial x_j^2} + 2\sum_{i<j}\delta_i\delta_j\frac{\partial^2 f(a)}{\partial x_i\partial x_j}\right] = \frac{1}{2!}\delta' A\delta, \qquad (1.3.2)$$

where

$$A = \frac{\partial}{\partial X}\frac{\partial}{\partial X'}f\Big|_{X=a} = \left[\begin{array}{cccc} \dfrac{\partial^2 f}{\partial x_1^2} & \dfrac{\partial^2 f}{\partial x_1\partial x_2} & \cdots & \dfrac{\partial^2 f}{\partial x_1\partial x_k} \\ \vdots & \vdots & \cdots & \vdots \\ \dfrac{\partial^2 f}{\partial x_k\partial x_1} & \dfrac{\partial^2 f}{\partial x_k\partial x_2} & \cdots & \dfrac{\partial^2 f}{\partial x_k^2} \end{array}\right]_{X=a}.$$

Note that $\delta' A\delta$ is a quadratic form in the matrix $A = A'$ of second-order partial derivatives, evaluated at a critical point $X = a$, where δ represents a small neighborhood of the point $X = a$. In this neighborhood if the quadratic form $\delta' A\delta > 0$ for all $\delta \neq O$ then it means that $f(a + \delta) - f(a) > 0$ or a neighborhood ordinate is bigger than the ordinate at $x = a$ or the point $X = a$ corresponds to a local minimum. Similarly if $\delta' A\delta < 0$ for all $\delta \neq O$ then in the neighborhood of the point $X = a$ the ordinate $f(a + \delta) < f(a)$ which means that $X = a$ corresponds to a local maximum. Therefore if A is positive definite then the point $X = a$ corresponds to a local minimum and if A is negative definite then the point $X = a$ corresponds to a local maximum. If neither of the above, then the point $X = a$ will be called a saddle point. Then whatever be the dimension k we can write the general expansion in the neighborhood of any point $X = a$ as follows:

$$f(a + \delta) = e^{\delta'\frac{\partial}{\partial X}}f|_{X=a} \quad \text{for } k = 1, 2, \ldots. \qquad (1.3.3)$$

Let

$$A = \frac{\partial}{\partial X}\frac{\partial}{\partial X'}f\Big|_{X=a}.$$

Then the general rule, irrespective of k, is the following: If $A > O$ then $X = a$ corresponds to a local minimum. If $A < O$ then the point $X = a$ corresponds to a local maximum. If A is indefinite or semidefinite then $X = a$ is a saddle point.

1.3.1.1. *Some special cases*

Case $k = 1$. The expansion is

$$f(a + \delta) = f(a) + \frac{1}{1!}f^{(1)}(a) + \frac{1}{2!}f^{(2)}(a) + \cdots,$$

$$f^{(r)}(a) = \frac{d^r}{dx^r}f(x)|_{x=a}.$$

If $f^{(2)}(a) > 0$ then $x = a$ corresponds to a minimum and if $f^{(2)}(a) < 0$ then $x = a$ corresponds to a maximum.

Case $k = 2$. The expansion is

$$f(a + \delta) = f(a_1 + \delta_1, a_2 + \delta_2) = f(a_1, a_2)$$
$$+ \frac{1}{1!}\left[\delta_1 \frac{\partial}{\partial x_1}f(a) + \delta_2 \frac{\partial}{\partial x_2}f(a)\right]$$
$$+ \frac{1}{2!}\left[\delta_1^2 \frac{\partial^2 f(a)}{\partial x_1^2} + \delta_2^2 \frac{\partial^2 f(a)}{\partial x_2^2} + 2\delta_1\delta_2 \frac{\partial^2 f(a)}{\partial x_1 \partial x_2}\right] + \cdots.$$

Hence the conditions are based on $A = (a_{ij})$, $a_{11} = \frac{\partial^2 f(a)}{\partial x_1^2}$, $a_{22} = \frac{\partial^2 f(a)}{\partial x_2^2}$, $a_{12} = \frac{\partial^2 f(a)}{\partial x_1 \partial x_2}$. $A > O$ (positive definite) means $a_{11} > 0, a_{22} > 0, a_{11}a_{22} - a_{12}^2 > 0$ then $X = a$ corresponds to a minimum. $A < O$ means $a_{11} < 0, a_{22} < 0, a_{11}a_{22} - a_{12}^2 > 0$ then $X = a$ corresponds to a maximum. If the above two cases are not there then the point corresponds to a saddle point.

Case $k = 3$. In this case the general expansion remains the same. The explicit forms will be the following:

$$f(a + \delta) = f(a) + \frac{1}{1!}\left[\sum_{j=1}^{3}\delta_j \frac{\partial f(a)}{\partial x_j}\right]$$

$$+ \frac{1}{2!}\left[\sum_{j=1}^{3}\delta_j^2 \frac{\partial^2 f(a)}{\partial x_j^2} + 2\sum_{i<j=1}^{3}\delta_i\delta_j \frac{\partial^2 f(a)}{\partial x_i \partial x_j}\right] + \cdots,$$

where A in this case is $A = \frac{\partial}{\partial X} \frac{\partial}{\partial X'} f|_{X=a}$. Let

$$
A = \begin{bmatrix} a_{11} & a_{12} & a_{13} \\ a_{12} & a_{22} & a_{23} \\ a_{13} & a_{23} & a_{33} \end{bmatrix}, \quad |A_{11}| = a_{11}, \quad |A_{22}| = \begin{vmatrix} a_{11} & a_{12} \\ a_{12} & a_{22} \end{vmatrix},
$$

$$
|A_{33}| = |A|.
$$

Then $A > O$ means $|A_{11}| > 0, |A_{22}| > 0, |A| > 0$ then the point $X = a$ corresponds to a minimum. $A < O$ means $|A_{11}| = a_{11} < 0, |A_{22}| > 0, |A| < 0$, then the point $X = a$ corresponds to a maximum. Otherwise the point corresponds to a saddle point.

Example 1.3.1. Expand $f = (1 - x_1 - 2x_2)^{-\frac{1}{2}}$ around the point $(x_1, x_2) = (0, 0)$ and check for convergence of the series.

Solution 1.3.1. Let $X = \begin{bmatrix} x_1 \\ x_2 \end{bmatrix}, \frac{\partial}{\partial X} = \begin{bmatrix} \frac{\partial}{\partial x_1} \\ \frac{\partial}{\partial x_2} \end{bmatrix}$ and consider the operator $e^{X' \frac{\partial}{\partial X}}$ operating on f and evaluated at $X = O$:

$$
\frac{\partial}{\partial x_1} [1 - x_1 - 2x_2]^{-\frac{1}{2}}|_{X=O}
$$

$$
= \left(-\frac{1}{2} \right) [1 - x_1 - 2x_2]^{-\frac{1}{2}-1}(-1)|_{X=O} = \frac{1}{2}.
$$

$$
\frac{\partial}{\partial x_2} [1 - x_1 - 2x_2]^{-\frac{1}{2}}|_{X=O}
$$

$$
= \left(-\frac{1}{2} \right) (-2)[1 - x_1 - 2x_2]^{-\frac{1}{2}-1}|_{X=O} = \frac{1}{2}(2).
$$

$$
\frac{\partial^2}{\partial x_1 \partial x_2} [1 - x_1 - 2x_2]^{-\frac{1}{2}}|_{X=O}
$$

$$
= \left(-\frac{1}{2} \right) \left(-\frac{3}{2} \right) (-1)(-2)[1 - x_1 - 2x_2]^{-\frac{5}{2}}|_{X=O}
$$

$$
= \frac{1}{2} \left(\frac{3}{2} \right) (-1)(-2),
$$

$$\frac{\partial^2}{\partial x_1^2}[1 - x_1 - 2x_2]^{-\frac{1}{2}}|_{X=O}$$

$$= \left(-\frac{1}{2}\right)\left(-\frac{3}{2}\right)(-1)^2[1 - x_1 - 2x_2]^{-\frac{5}{2}}|_{X=O}$$

$$= \left(\frac{1}{2}\right)\left(\frac{3}{2}\right)(-1)^2,$$

$$\frac{\partial^2}{\partial x_2^2}[1 - x_1 - 2x_2]^{-\frac{1}{2}}|_{X=O}$$

$$= \left(-\frac{1}{2}\right)(-2)\left(-\frac{3}{2}\right)(-2)[1 - x_1 - 2x_2]^{-\frac{5}{2}}|_{X=O}$$

$$= \left(\frac{1}{2}\right)\left(\frac{3}{2}\right)(-2)^2.$$

Hence

$$(1 - x_1 - 2x_2)^{-\frac{1}{2}} = e^{X'\frac{\partial}{\partial X}}f|_{X=O} = 1 + \frac{\left(\frac{1}{2}\right)}{1!}[x_1 + 2x_2]$$

$$+ \frac{\left(\frac{1}{2}\right)\left(\frac{3}{2}\right)}{2!}[x_1^2 + 2x_1 x_2 + 4x_2^2] + \cdots.$$

For the convergence of the series we can check the general term or from the starting binomial expansion. The condition is $|x_1 + 2x_2| < 1$. We may also observe that the final expansion is of the form $(1 - u)^{-\frac{1}{2}}$ for $|u| < 1$ with $u = x_1 + 2x_2$.

Example 1.3.2. Expand the function in Example 1.3.1 around the point $(x_1, x_2) = (\frac{1}{5}, \frac{1}{5})$.

Solution 1.3.2. We may consider the operator $\exp\{(x_1 - \frac{1}{5})\frac{\partial}{\partial x_1} + (x_2 - \frac{1}{5})\frac{\partial}{\partial x_2}\}$ operating on $f = (1 - x_1 - 2x_2)^{-\frac{1}{2}}$ and evaluated at $(x_1, x_2) = (\frac{1}{5}, \frac{1}{5})$ or we may replace x_j by $\frac{1}{5} + \delta_j, j = 1, 2$ and expand for (δ_1, δ_2) around $(0, 0)$ we will get the same result.

Moreover,

$$f(x_1, x_2)|_{(\frac{1}{5}, \frac{1}{5})} = \left(1 - \frac{1}{5} - \frac{2}{5}\right)^{-\frac{1}{2}} = \left(\frac{2}{5}\right)^{-\frac{1}{2}} = \frac{\sqrt{5}}{\sqrt{2}},$$

$$\frac{\partial f}{\partial x_1}\Big|_{(\frac{1}{5},\frac{1}{5})} = \left(-\frac{1}{2}\right)(-1)(1 - x_1 - 2x_2)^{-\frac{3}{2}}\big|_{(\frac{1}{5},\frac{1}{5})} = \left(\frac{1}{2}\right)\left(\frac{5}{2}\right)^{\frac{3}{2}},$$

$$\frac{\partial f}{\partial x_2}\Big|_{(\frac{1}{5},\frac{1}{5})} = \left(-\frac{1}{2}\right)(-2)(1 - x_1 - 2x_2)^{-\frac{3}{2}} = \left(\frac{5}{2}\right)^{\frac{3}{2}}.$$

Similarly, we compute the second-order derivatives and evaluate at $(\frac{1}{5}, \frac{1}{5})$ to obtain the following:

$$(1 - x_1 - 2x_2)^{-\frac{1}{2}} = \frac{\sqrt{5}}{\sqrt{2}} + \frac{\left(\frac{1}{2}\right)}{1!}\left(\frac{5}{2}\right)^{\frac{3}{2}}\left[\left(x_1 - \frac{1}{5}\right) + 2\left(x_2 - \frac{1}{5}\right)\right]$$

$$+ \frac{\left(\frac{1}{2}\right)\left(\frac{3}{2}\right)}{2!}\left(\frac{5}{2}\right)^{\frac{5}{2}}\left[\left(x_1 - \frac{1}{5}\right)^2\right.$$

$$\left. + 2\left(x_1 - \frac{1}{5}\right)\left(x_2 - \frac{1}{5}\right) + 4\left(x_2 - \frac{1}{5}\right)^2\right] + \cdots.$$

Exercise 1.3

1.3.1. Expand into power series (i): e^x around $x = 0$, (ii): e^x around $x = 2$, (iii): e^x around $x = -1$.

1.3.2. Expand into power series $\sin x$ around (i): $x = 0$, (ii): $x = \frac{\pi}{4}$, (iii): $x = -\frac{\pi}{2}$.

1.3.3. Expand into power series (a): $\sin(x_1 + x_2)$ around (i): $(x_1, x_2) = (0, 0)$, (ii): $(x_1, x_2) = (\frac{\pi}{4}, \frac{\pi}{4})$, (iii): $(x_1, x_2) = (\frac{\pi}{4}, \frac{\pi}{2})$, (iv): $(x_1, x_2) = (-\frac{\pi}{4}, \frac{\pi}{4})$, (b): $\cos(x_1 + x_2)$ around the same points as in (a) above.

1.3.4. Expand into power series $\exp\{x_1 + 2x_2 - x_3\}$ around (i): $(0, 0, 0)$, (ii): $(1, 0, -1)$ and work out the conditions for convergence of the series in each case.

1.3.5. Expand into power series $(1 - 2x_1 - 3x_2)^{-\frac{2}{3}}$ around (i): $(0, 0)$, (ii): $(\frac{1}{7}, \frac{1}{7})$ and check for convergence in each case.

1.3.6. Expand into power series $\ln(1 + x_1 + 2x_2)$ around (i): $(0, 0)$, (ii): $(\frac{1}{3}, \frac{1}{3})$ and check for convergence in each case.

1.3.7. Expand into power series $(x_1 - x_1^2 x_2 - x_1 x_2)^{\frac{5}{2}}$ around (i): $(0, 0)$, (ii): $(-\frac{1}{2}, \frac{1}{2})$ and check for convergence in each case.

1.3.8 Expand into power series $(1 - a_1x_1 - a_2x_2)^{-\alpha}$ where a_1 and a_2 are constants, around (i): $(x_1, x_2) = (0,0)$, (ii): $(x_1, x_2) = (b_1, b_2)$ and give the conditions for convergence in each case.

1.3.9. Expand into power series $(1 - a_1x_1 - a_2x_2 - a_3x_3)^{-\alpha}$, where a_1, a_2, a_3 are constants, around (i): $(x_1, x_2, x_3) = (0,0,0)$, (ii): (b_1, b_2, b_3) and give the conditions for convergence of the series in each case.

1.3.10. Repeat Exercise 1.3.9 for the general case where there are k terms $a_1x_1 + \cdots + a_kx_k$ and expand around the points (i): $(0, \ldots, 0)$, (ii): $(b_1, \ldots, b_k)$.

Example 1.3.3. Consider

$$f(X) = X'AX, \quad X = \begin{bmatrix} x_1 \\ x_2 \\ x_3 \end{bmatrix}, \quad A = \begin{bmatrix} 2 & -1 & 0 \\ -1 & 3 & 2 \\ 0 & 2 & 5 \end{bmatrix}$$

or $f(x_1, x_2, x_3) = 2x_1^2 + 3x_2^2 + 5x_3^2 - 2x_1x_2 + 4x_2x_3$. Check for maxima/minima.

Solution 1.3.3. $\frac{\partial f}{\partial X} = O \Rightarrow 2AX = O \Rightarrow X = O$ since A is non-singular. There is only one critical point $X = O$. Now, $\frac{\partial}{\partial X}\frac{\partial}{\partial X'}f = 2A$. The leading minors of A are the following:

$$2 > 0, \quad \begin{vmatrix} 2 & -1 \\ -1 & 3 \end{vmatrix} = 5 > 0, \quad |A| = 17 > 0$$

and hence $A > O$ or A is positive definite and the critical point corresponds to a minimum.

Example 1.3.4. Consider

$$f(X) = X'AX + X'b + 8, \quad X = \begin{bmatrix} x_1 \\ x_2 \\ x_3 \end{bmatrix},$$

$$b = \begin{bmatrix} 5 \\ -1 \\ 2 \end{bmatrix}, \quad A = \begin{bmatrix} 2 & -1 & 0 \\ -1 & 3 & 2 \\ 0 & 2 & 5 \end{bmatrix}$$

or $f(x_1, x_2, x_3) = 2x_1^2 + 3x_2^2 + 5x_3^2 - 2x_1x_2 + 4x_2x_3 + 5x_1 - x_2 + 2x_3 + 8$.
Check for maxima/minima.

Solution 1.3.4. Proceeding as in the solution of Example 1.3.3 we
have $\frac{\partial f(X)}{\partial X} = 2AX + b = O$. This equation has only one solution and
therefore only one critical point. The solution is $X = -(\frac{1}{2}A^{-1}b)$. But
we need not solve for the critical point since there is only one critical
point. From Example 1.3.3 we have $\frac{\partial}{\partial X} \frac{\partial}{\partial X'} f(X) = 2A$. We have
already seen that this A is positive definite and hence the critical
point corresponds to a minimum.

A positive definite quadratic form in the real case, equated to
a positive constant, such as $X'AX = c, c > 0$ is the surface of an
ellipsoid in general. If the number of variables is k and if $k = 2$
then we have an ellipse in a plane. If the x_1x_2 term is absent then
the ellipse is in the standard form and if not it is an offset ellipse
or obtained by rotating the axes through an angle. For a general k,
consider $f = X'AX + X'b = c$ where if the $k \times k$ matrix $A > O$, if
the $k \times 1$ vector b and the 1×1 scalar c are such that we can write
$f = (X - \mu)'A(X - \mu) = \alpha, \alpha > 0$ for some μ and α, then we have
a relocated offset ellipsoid. For arbitrary α this will be unbounded.
If we can confine this ellipsoid within a hypersphere or something
like that, then we can talk about a finite maximum. We will consider
such problems in the following section.

1.3.2. Maxima/minima subject to constraints

Let us take a simple problem of optimizing a quadratic function
subject to quadratic constraint. Optimize $X'AX, A = A'$ subject
to the condition $X'X = 1$. We want to look for maxima/minima
when the quadratic form is confined to a hypersphere of radius 1.
If $X'X = 1$ then $X'X - 1 = 0$, that is, $\lambda(X'X - 1) = 0$ where λ
is an arbitrary constant. Let us add zero to $X'AX$ and consider the

function

$$f = X'AX - \lambda(X'X - 1) \qquad (a)$$

which is nothing but $X'AX$ because we added only a constant multiple of zero. This method is called the method of Lagrangian multipliers. Here there is one Lagrangian multiplier λ. If there is a local maximum or local minimum then $\frac{\partial f}{\partial X} = O$ and taking λ as an additional variable, $\frac{\partial f}{\partial \lambda} = 0$ also. That is,

$$\frac{\partial f}{\partial X} = O \Rightarrow 2AX - 2\lambda X = O \Rightarrow AX = \lambda X. \qquad (b)$$

The second equation gives back the constraint $X'X - 1 = 0$. If (b) has a non-null X as a solution then the coefficient matrix $A - \lambda I$ has to be singular or $|A - \lambda I| = 0$.

$$|A - \lambda I| = 0 \quad \text{and} \quad X'X = 1. \qquad (1.3.4)$$

This shows that λ is an eigenvalue of A. Premultiplying (b) by X' and using the condition $X'X = 1$ we have

$$AX = \lambda X \Rightarrow X'AX = \lambda X'X = \lambda. \qquad (1.3.5)$$

Hence the maximum value of $X'AX$ is the largest eigenvalue of A and the minimum value is the smallest eigenvalue of A. This may be stated as a result.

Result 1.3.1. $\max_{X'X=1} X'AX = \lambda_1$ *and* $\min_{X'X=1} X'AX = \lambda_p$, *where* λ_1 *is the largest eigenvalue and* λ_p *the smallest eigenvalue of the* $p \times p$ *matrix* A.

In the above case we were successful in finding the maximum and minimum by studying the eigenvalues of A. Is it possible to check for maxima/minima by our usual method of finding the matrix of second-order derivatives and checking for definiteness of this matrix of second-order derivatives? When Lagrangian multipliers are there we have all the original variables X and the Lagrangian multipliers as the new variables. In this case our variables are X and λ. The matrix of second-order derivatives is available from the following submatrices. $\frac{\partial}{\partial X}\frac{\partial f}{\partial X'}$, $\frac{\partial}{\partial \lambda}\frac{\partial f}{\partial X}$, $\frac{\partial}{\partial \lambda}\frac{\partial f}{\partial X'}$, $\frac{\partial^2 f}{\partial \lambda^2}$. Then we have the following matrix of second-order derivatives:

$$\begin{bmatrix} 2(A - \lambda_0 I) & 2X_0 \\ 2X_0' & 0 \end{bmatrix} = 2 \begin{bmatrix} A - \lambda_0 I & X_0 \\ X_0' & 0 \end{bmatrix} = B \text{ (say)},$$

where X_0 is the eigenvector corresponding to the eigenvalue λ_0. This matrix is evidently indefinite because one diagonal element is already zero. If $B > O$ then all diagonal elements of B must be positive and if $A < O$ then all diagonal elements of B must be negative. Here B is indefinite. In general, if the method of Lagrangian multipliers is used then the matrix of second-order derivatives will always be indefinite. Suppose that we ignore the Lagrangian multiplier λ and take only the second-order derivatives in the original variables X. Then the matrix is $2(A - \lambda I)$ which is obviously singular and not positive or negative definite. Hence this also fails. When the method of Lagrangian multipliers is used we can arrive at the critical points easily but checking for maxima/minima at these critical points has to be done through some other means.

1.3.3. Optimization of a quadratic form subject to general quadratic constraint

Consider the optimization of $X'AX$ subject to $X'BX = 1$ where $A = A', B = B'$ are positive definite $p \times p$ constant matrices and X is unknown. This problem is a generalization of the previous problem of optimizing $X'AX$ subject to $X'X = 1$. Let

$$f = X'AX - \lambda(X'BX - 1).$$

Then

$$\frac{\partial f}{\partial X} = O \Rightarrow AX = \lambda BX \Rightarrow X'AX = \lambda X'BX = \lambda.$$

Note that $AX = \lambda BX$ also means $(B^{-1}A - \lambda I)X = O$ or λ is an eigenvalue of $B^{-1}A$. Hence we have the following result.

Result 1.3.2.

$$\max_{X'BX=1} X'AX = \lambda_1 \quad and \quad \min_{X'BX=1} X'AX = \lambda_p, \qquad (1.3.6)$$

where λ_1 is the largest and λ_p is the smallest eigenvalue of $B^{-1}A$.

Note that the eigenvector, corresponding to an eigenvalue λ_j of $B^{-1}A$ has to be obtained through $X'BX = 1$. After taking an eigenvector corresponding to λ_j from the equation $(B^{-1}A - \lambda_j I)X_j = O$ compute $X'_j BX_j = \alpha$ say. Then $U_j = \frac{1}{\sqrt{\alpha}} X_j$ is the eigenvector satisfying the condition $U'_j BU_j = 1$.

1.3.3.1. *Principal components analysis*

There is a very popular practical problem connected with optimization of a positive definite quadratic form subject to a quadratic constraint of the type $X'X = 1$. This is known as Principal Components Analysis in the area of data analysis.

Suppose that a medical doctor decided to come up with a formula for "good health" among 20-year old girls. He decided to take observations on a number of possible variables, such as x_1 = height, x_2 = weight, x_3 = average amount of food consumed daily, etc. She has taken observations on 1000 variables. Now, she does not know what to do with all these observations on these 1000 variables. Since there are too many variables she has to decide which variables are important for her study so that other variables can be deleted. How do we decide which variables are important and which variables are unimportant? What should be a criterion by which one can say certain variables are to be included in our study and other variables can be safely neglected? If the 20-year old girls are from the same genetical group of the same community then their heights may be more or less the same, say 5 feet. Hence we already know that x_1 should be more or less 5 or it is predetermined as far as this study is concerned and hence x_1 is not an important variable to be included in the study because we already know it to be 5. But x_2 = weight may have considerable variation among the 20-year old. Hence this x_2 is to be included in our study. Thus a good criterion that can be adopted for any such study is that variables having large variation or spread or scatter must be included in the study. Instead of considering individual variables we can take linear functions of the variables because linear functions also contain individual variables. Let $u = a_1x_1 + \cdots + a_kx_k$ be an arbitrary linear function of the variables $x_1, \ldots, x_k$ where $a_1, \ldots, a_k$ are constants. A good measure of spread in the variable u is the standard deviation in u or the square of the spread or dispersion is the variance, denoted by $\mathrm{Var}(u) = E(u - E(u))^2$ where E denotes the expected value. It can be shown that this variance has the formula

$$\mathrm{Var}(u) = a'Va$$

$$a = \begin{bmatrix} a_1 \\ \vdots \\ a_k \end{bmatrix}, \quad V = (v_{ij}), \quad v_{jj} = \mathrm{Var}(x_j), \quad v_{ij} = \mathrm{Cov}(x_i, x_j),$$

$$(1.3.7)$$

where $\mathrm{Cov}(x_i, x_j)$ = covariance between x_i and x_j or the joint scatter in (x_i, x_j). This V is usually called the covariance matrix or the variance–covariance matrix. We would like to select the coefficient vector a such that the variance $a'Va$ is maximized. But for unbounded a the maximum is at infinity and hence we may assume that the sum of squares of the coefficients is 1 or $a_1^2 + \cdots + a_k^2 = 1$ or $a'a = 1$. Hence our problem reduces to maximizing $a'Va$ subject to $a'a = 1$. We already know the solution.

$$\max_{a'a=1} a'Va = \lambda_1,$$

where λ_1 is the largest eigenvalue of V. In order to compute λ_1 we must know V. In a practical situation, usually V is unknown. In this case what is usually done is to have an estimate of V and use that estimate for computational purposes. Let x_{ir} denote the rth observation on x_i and suppose that n observations each are taken on each variable. Then v_{ij} is estimated by

$$\hat{v}_{ij} = \frac{\sum_{r=1}^{n}(x_{ir} - \bar{x}_i)(x_{jr} - \bar{x}_j)}{n},$$

$$\hat{v}_{jj} = \frac{\sum_{r=1}^{n}(x_{jr} - \bar{x}_j)^2}{n},$$

where $\bar{x}_j = \frac{\sum_{r=1}^{n} x_{jr}}{n}$ and $\hat{V} = (\hat{v}_{ij})$. This $\hat{V}$ is used instead of V when V is unknown. Let λ_1 be the largest eigenvalue of V or $\hat{V}$ as the case may be. Let the corresponding eigenvector, normalized through $a'a = 1$, be $a_{(1)}$. Then $u_1 = a'_{(1)}X$ is the most important linear function or called the first principal component. Now, start with the second largest eigenvalue of V, say λ_2 and let the corresponding normalized (normalized through $a'a = 1$) eigenvector be $a_{(2)}$. Then $u_2 = a'_{(2)}X$ is the second principal component. Continue the process. Observe that $\text{Var}(u_j) = \lambda_j$. Stop the process when λ_j falls below a pre-selected number or select the eigenvalues greater than or equal to this pre-selected number. This means our tolerance level is that we will allow variance as small as the pre-selected number. Thus, the mathematical part of Principal Components Analysis is a very simple problem of maximizing a positive definite quadratic form subject to a simple quadratic constraint. But interpretations of linear functions, which are the principal components, may be difficult in a practical situation.

Example 1.3.5. This method is used only when we have a large number of variables. For the sake of illustration of the steps we will illustrate on the following covariance matrix A.

Solution 1.3.5. Let

$$A = \begin{bmatrix} 2 & -\sqrt{2} & 0 \\ -\sqrt{2} & 3 & 0 \\ 0 & 0 & 1 \end{bmatrix}, \quad |A - \lambda I| = 0 \Rightarrow \lambda_1 = 4, \lambda_2 = 1, \lambda_3 = 1.$$

Consider $\lambda_1 = 4$.

$$(A - \lambda_1 I)a = O \Rightarrow \begin{bmatrix} -2 & -\sqrt{2} & 0 \\ -\sqrt{2} & -1 & 0 \\ 0 & 0 & -3 \end{bmatrix} \begin{bmatrix} a_1 \\ a_2 \\ a_3 \end{bmatrix}$$

$$= \begin{bmatrix} 0 \\ 0 \\ 0 \end{bmatrix} \Rightarrow a = \begin{bmatrix} 1 \\ -\sqrt{2} \\ 0 \end{bmatrix}.$$

Normalizing through $a'a = 1$ we have $a'a = 3 \Rightarrow$ the normalized a, denoted by $a_{(1)}$ is such that $a'_{(1)} = (\frac{1}{\sqrt{3}}[1, -\sqrt{2}, 0])$. Hence the first principal component is

$$u_1 = \frac{x_1}{\sqrt{3}} - \frac{\sqrt{2}x_2}{\sqrt{3}}.$$

Note that

$$\text{Var}(u_1) = \tfrac{1}{3}\text{Var}(x_1) + \tfrac{2}{3}\text{Var}(x_2) - 2 \times \tfrac{1}{\sqrt{2}} \times \tfrac{\sqrt{2}}{\sqrt{3}}\text{Cov}(x_1, x_2)$$

$$= \tfrac{1}{3}(2) + \tfrac{2}{3}(3) - \tfrac{2\sqrt{2}}{3}(-\sqrt{2}) = 4.$$

Now, consider $\lambda_2 = 1$

$$(A - \lambda_2 I)a = O \Rightarrow \begin{bmatrix} 1 & -\sqrt{2} & 0 \\ -\sqrt{2} & 2 & 0 \\ 0 & 0 & 0 \end{bmatrix} \begin{bmatrix} a_1 \\ a_2 \\ a_3 \end{bmatrix} = \begin{bmatrix} 0 \\ 0 \\ 0 \end{bmatrix} \Rightarrow a = \begin{bmatrix} \sqrt{2} \\ 1 \\ 0 \end{bmatrix}.$$

Normalizing through $a'a = 1$ we have $a'_{(2)} = \frac{1}{\sqrt{3}}[\sqrt{2}, 1, 0]$. Another solution is $a'_{(3)} = [0, 0, 1]$. Hence the second and third principal components are

$$u_2 = \frac{\sqrt{2}}{\sqrt{3}}x_1 + \frac{1}{\sqrt{3}}x_2, u_3 = x_3.$$

Note that

$$\text{Var}(u_2) = \tfrac{2}{3}\text{Var}(x_1) + \tfrac{1}{3}\text{Var}(x_2) + 2 \times \tfrac{\sqrt{2}}{\sqrt{3}} \times \tfrac{1}{\sqrt{3}}\text{Cov}(x_1, x_2)$$
$$= \tfrac{2}{3}(2) + \tfrac{1}{3}(3) - \tfrac{2}{3}(2) = 1$$

and $\text{Var}(u_3) = \text{Var}(x_3) = 1$. Here u_1, u_2, u_3 are the principal components in this case. As indicated earlier, this example is for illustration only. We apply this method only when the number of variables is large so that the number of linear functions (principal components) will be significantly smaller than the number of original variables.

1.3.3.2. *Maximizing a quadratic form subject to linear constraint*

This problem is of the type of maximizing $X'AX, A = A'$ subject to $X'b = c$ where the $p \times p$ matrix A, the $p \times 1$ vector b and the 1×1 matrix or scalar are known. By using $X'b = c$ we can eliminate one variable, substitute that in $X'AX$ and then maximize the resulting quadratic expression. This is one way of doing the problem. Another way is to use Lagrangian multiplier. Consider

$$f = X'AX - 2\lambda(X'b - c). \tag{1.3.8}$$

Here the Lagrangian multiplier -2λ is taken for convenience only. Then the normal equations (the equations which give the critical points) are the following:

$$\frac{\partial f}{\partial X} = O \Rightarrow 2AX = 2\lambda b. \tag{i}$$

That is,

$$X = \lambda A^{-1}b \Rightarrow c = X'b = \lambda b' A^{-1}b \text{ or } \lambda = \frac{c}{b'A^{-1}b}. \tag{ii}$$

From (i),

$$X'AX = \lambda X'b = \lambda c = \frac{c^2}{b'A^{-1}b} \tag{1.3.9}$$

is the largest value of $X'AX$.

1.3.4. Maximizing a linear function subject to quadratic constraint

Consider arbitrary linear function $X'a$ where a is a known $p \times 1$ vector and consider the constraint $X'AX = 1, A = A'$, where A is known. We want to maximize $X'a$ subject to $X'AX = 1$. A practical problem in elementary calculus is to maximize the sum $x_1 + \cdots + x_p$, given that the sum of squares is fixed, say $x_1^2 + \cdots + x_p^2 = 1$. If $x_1, \ldots, x_p$ are continuous then the problem is simple but if $x_1, \ldots, x_p$ can only take integer values then the problem is complicated which falls in the category of integer programming problem, a solution of which may be seen from [8]. Let the variables be continuous and let $\frac{\lambda}{2}$ be a Lagrangian multiplier and let us consider

$$f = X'a - \frac{\lambda}{2}(X'AX - 1), \quad A = A', \quad A > O.$$

Then

$$\frac{\partial f}{\partial X} = O \Rightarrow a - \lambda AX = O \Rightarrow a = \lambda AX$$

and $X'a = \lambda X'AX = \lambda$. Hence the largest value of $X'A$ is the largest value of λ. Also from $a = \lambda AX$ we have $A^{-1}a = \lambda X \Rightarrow a'A^{-1}a = \lambda a'X = \lambda^2$ or $\lambda = \sqrt{a'A^{-1}a}$.

This can also be obtained from a simple argument by using the Cauchy–Schwartz inequality. Consider

$$a'X = a'A^{-\frac{1}{2}}A^{\frac{1}{2}}X = (A^{-\frac{1}{2}}a)'(A^{\frac{1}{2}}X)$$

$$\leq \sqrt{a'A^{-1}a}\sqrt{X'AX} = \sqrt{a'A^{-1}a} \qquad (1.3.10)$$

since $X'AX = 1$. The above inequality is Cauchy–Schwartz's inequality and $A^{\frac{1}{2}}$ denotes the symmetric positive definite square root of A. Note that for taking A^{-1} one needs only A to be non-singular but for taking a unique square root one needs A to be positive definite. Also, for applying Cauchy–Schwartz inequality one needs $a'A^{-1}a$ to remain positive. There are several applications of the result (1.3.10) in statistical theory of estimation.

1.3.5. Maximizing a bilinear form subject to quadratic constraint

In Section 1.3.4 we considered optimizing a linear form subject to quadratic constraint. The present one is a generalization of that problem. A bilinear form is of the following type: $X'AY$ where

$$
X = \begin{bmatrix} x_1 \\ \vdots \\ x_p \end{bmatrix}, \quad
Y = \begin{bmatrix} y_1 \\ \vdots \\ y_q \end{bmatrix}, \quad
A = \begin{bmatrix} a_{11} & \cdots & a_{1q} \\ \vdots & \cdots & \vdots \\ a_{p1} & \cdots & a_{pq} \end{bmatrix}, \quad X'AY.
$$

That is, X is $p \times 1$, Y is $q \times 1$ and A is $p \times q$ where p may or may not be equal to q. Note that $X'AY$ is linear in X as well as linear in Y and hence called a bilinear form. Also, $X'AY$ is 1×1 and hence it is equal to its transpose or $X'AY = Y'A'X$. Consider

$$
x_1 y_1 + 2x_2 y_1 - 2x_3 y_1 - x_1 y_2 + 5x_2 y_2 + x_3 y_2 = X'AY = Y'A'X,
$$

where

$$
X = \begin{bmatrix} x_1 \\ x_2 \\ x_3 \end{bmatrix}, \quad
Y = \begin{bmatrix} y_1 \\ y_2 \end{bmatrix}, \quad
A = \begin{bmatrix} 1 & -1 \\ 2 & 5 \\ -2 & 1 \end{bmatrix}.
$$

Let the constraints be $X'BX = 1$ and $Y'CY = 1$ where $B > O$ and $C > O$ are known positive definite matrices. Consider the Lagrangian multipliers $\frac{\lambda_1}{2}$ and $\frac{\lambda_2}{2}$ and consider the function

$$
f = X'AY - \frac{\lambda_1}{2}(X'BX - 1) - \frac{\lambda_2}{2}(Y'CY - 1).
$$

The critical points are available from the equations $\frac{\partial f}{\partial X} = O$ and $\frac{\partial f}{\partial Y} = O$. Since $X'AY = Y'A'X$, when differentiating with respect to Y we may use the form $Y'A'X$ for convenience. That is,

(i) $\quad \dfrac{\partial f}{\partial X} = O \Rightarrow AY = \lambda_1 BX,$

(ii) $\quad \dfrac{\partial f}{\partial Y} = O \Rightarrow A'X = \lambda_2 CY.$

Premultiply (i) X' and (ii) by Y'. Then

(iii) $X'AY = \lambda_1 X'BX = \lambda_1$ and $Y'A'X = \lambda_2 Y'CY = \lambda_2$.

This means $\lambda_1 = \lambda_2 = \lambda$ (say). Then (i) and (ii) can be rewritten as

(i) $AY - \lambda BX = O \Rightarrow \dfrac{1}{\lambda} B^{-1} AY = X,$

(ii) $-\lambda CY + A'X = O \Rightarrow \lambda Y = C^{-1} A'X.$

Taking X from (i) and substituting in (ii) we have

(iv) $(C^{-1} A' B^{-1} A - \lambda^2 I)Y = O$

and

(v) $(B^{-1} A C^{-1} A' - \lambda^2 I)X = O.$

From (iv) and (v), it follows that λ^2 is an eigenvalue of $G_2 = C^{-1} A' B^{-1} A$ as well as $G_1 = B^{-1} A C^{-1} A'$. Hence when evaluating the eigenvalues we may use any of these matrices. If $p < q$ then use $G_1 = B^{-1} A C^{-1} A'$ because this is $p \times p$, otherwise use $G_2 = C^{-1} A' B^{-1} A$ which is $q \times q$. The eigenvector X corresponding to an eigenvalue λ^2 is available from G_1 and the eigenvector Y is available from G_2. But these X and Y are to be normalized through $X'BX = 1$ and $Y'CY = 1$. From (iii), $\lambda = X'AY = Y'A'X$ the maximum. Hence we have the following result.

Result 1.3.3.

$$\max_{X'BX=1, Y'CY=1} X'AY = \lambda_1 \quad \text{and} \quad \min_{X'AX=1, Y'CY=1} X'AY = \lambda_p,$$

$$(1.3.11)$$

where λ_1^2 is the largest eigenvalue of G_1 or G_2 and λ_p^2 is the smallest eigenvalue of G_1 or G_2.

A very popular application of this result is the famous canonical correlation analysis in the area of multivariate data analysis.

1.3.5.1. *Canonical correlation analysis*

This is a very popular technique of prediction in many disciplines. The basic idea is the following: Suppose that we wish to predict the increase in the yield of milk of a milking cow under an experimental feed x. We would like to answer questions such as what is the expected yield of milk y at a pre-assigned value of x such as if $x = 2$ kg (kilogram) of the special feed is given what is the expected yield of milk? What is the best prediction function of x for predicting y? We can take any arbitrary function such as a linear function $y = 3 + 0.5x$ and use it as a predictor function but the predicted value may be too far away from the observed value. For example if $x = 2$ then this prediction function gives $2 + (0.5)(2) = 3$ liters of milk but the actual milk obtained may be 5 liters. Hence the error in using the above prediction function, at $x = 2$, is $5 - 3 = 2$ liters. What is the best prediction function so that the error in prediction is minimized? Here y is a random variable and we can show that the best prediction function, best in the sense of minimizing a Euclidean distance, is the conditional expectation of y, at given x, denoted by $E(y|x)$. If y is to be predicted by using several real scalar variables $x_1, \ldots, x_k$, such as $x_1 =$ amount of special feed, $x_2 =$ amount of grass, $x_3 =$ amount of water, etc., then we can show that the best prediction function is $E(y|x_1, \ldots, x_k)$ or the conditional expectation of y at given values of $x_1, \ldots, x_k$. A general prediction problem of this category is the following: Suppose we have two sets of real scalar variables $\{x_1, \ldots, x_p\}$ and $\{y_1, \ldots, y_q\}$. Suppose that the variables in the y-set are to be predicted by using the variables in the x-set or vice versa. Instead of considering individual variables we may consider linear functions. Consider an arbitrary linear function $u = a_1 x_1 + \cdots + a_p x_p$ of $x_1, \ldots, x_p$ and let $v = b_1 y_1 + \cdots + b_q y_q$ be an arbitrary linear function of $y_1, \ldots, y_q$, where $a_1, \ldots, a_p, b_1, \ldots, b_q$ are constants. Then the variances of u and v will be quadratic forms of the type $a'Ba$ and $b'Cb$ respectively where $B = B' > O$ and $C = C' > O$ and the covariance between u and v will be a bilinear form of the type $a'Ab = b'A'a$. Then the problem of best prediction reduces to maximizing $a'Ab$ subject to the conditions $a'Ba =$ fixed, say 1 and $b'Cb =$ fixed, say 1 or to maximize $a'Ab$ subject to

$a'Ba = 1$ and $b'Cb = 1$ and come up with the best vectors a and b so that the best linear functions are $a'X$ and $b'Y$ to predict each other, $X' = (x_1, \ldots, x_p), Y' = (y_1, \ldots, y_q)$. We have already seen from Section 1.3.5 that the maximum of $a'Ab$ is λ_1 where λ_1^2 is the largest eigenvalue of $G_1 = B^{-1}AC^{-1}A'$ or $G_2 = C^{-1}A'B^{-1}A$. Then construct the eigenvector, corresponding to λ_1, through G_1 to get the best coefficient vector, a. Then normalize through $a'Ba = 1$. Such a normalized a be denoted by $a_{(1)}$. Similarly, construct the eigenvector corresponding to λ_1 through G_2 to get the best coefficient vector b. Normalize through $b'Cb = 1$, call this normalized vector $b_{(1)}$. Then $u_1 = a'_{(1)}X$ and $v_1 = b'_{(1)}Y$ is the first canonical pair of linear functions to predict each other. Now, start with the second largest eigenvalue λ_2 and construct the best normalized coefficient vectors $a_{(2)}$ and $b_{(2)}$ so that the second pair of canonical variables is $(u_2, v_2) = (a'_{(2)}X, b'_{(2)}Y)$, and so on. A numerical example will take too much space and hence it is left to the students to do the problems in the exercises.

Exercise 1.3 (*Continued*)

1.3.11. Evaluate the principal components if the covariance matrix is

$$\text{(i)} \quad V = \begin{bmatrix} 3 & -1 \\ -1 & 2 \end{bmatrix}, \quad \text{(ii)} \quad V = \begin{bmatrix} 2 & 1 & -1 \\ 0 & 1 & 4 \\ -1 & 1 & 5 \end{bmatrix}$$

and verify that the variances of the principal components are the eigenvalues of V in each case.

1.3.12. Check for the definiteness of the following matrices, if possible.

$$\text{(i)} \quad A_1 = \begin{bmatrix} 1 & -2 & 1 \\ 0 & 1 & 4 \\ 0 & 0 & 5 \end{bmatrix}, \quad \text{(ii)} \quad A_2 = \begin{bmatrix} 2 & -1 & 1 \\ -1 & 3 & 0 \\ 1 & 0 & -2 \end{bmatrix},$$

(iii) $A_3 = \begin{bmatrix} -2 & 1 & 0 \\ 1 & -3 & 1 \\ 0 & 1 & -2 \end{bmatrix}$, (iv) $A_4 = \begin{bmatrix} 2 & 2 & 0 \\ 2 & 2 & 1 \\ 0 & 1 & 4 \end{bmatrix}$.

1.3.13. Write the following in the form $X'AX + X'b + c, A = A'$ and then optimize:

(i) $u_1 = 2x_1^2 + 5x_2^2 + x_3^2 - 2x_1x_2 - x_1x_3 + 3x_1 + x_2 - 4x_3 + 5$,

(ii) $u_2 = x_1^2 - 2x_2^2 + 2x_3^2 - x_1x_2 + 2x_2x_3 + 4x_1 - x_2 + x_3 + 8$,

(iii) $u_3 = u_1$ above with the constraint $x_1 + 2x_2 + x_3 = 1$,

(iv) $u_4 = u_1$ above with the constraint $2x_1^2 + 3x_2^2 + x_3^2 = 1$,

(v) $u_5 = u_1$ above with the constraint $x_1 + x_2 + x_3 = 1$,

$$x_1^2 + 2x_2^2 + x_3^2 + x_1x_2 = 1.$$

1.3.14. Maximize $2x_1 + x_2 + 3x_3$ subject to $x_1^2 + x_2^2 + x_3^2 = 1$.

1.3.15. Maximize $x_1 + 2x_2 + x_3$ subject to $2x_1^2 + 3x_2^2 + 2x_1x_2 + x_3^2 = 1$.

1.3.16. Optimize $2x_1^2 + 4x_2^2 - 2x_1x_2 + 5x_3^2$ subject to $2x_1^2 + 3x_2^2 + 2x_1x_2 + x_3^2 = 1$.

1.3.17. Optimize $x_1y_1 + x_2y_1 - x_1y_2 + 2x_2y_2$ subject to $3x_1^2 + 5x_2^2 - 2x_1x_2 = 1, y_1^2 + y_2^2 = 1$.

1.3.18. Optimize $2x_1y_1 - x_1y_2 + x_2y_1 + x_2y_2$ subject to $2x_1^2 + x_2^2 - 2x_1x_2 = 1, y_1^2 + 3y_2^2 + 2y_1y_2 = 1$.

1.3.19. Optimize $x_1y_1 + x_1y_2 + x_1y_3 - x_2y_1 + x_2y_2 + 2x_2y_3$ subject to $2x_1^2 + 3x_2^2 - 2x_1x_2 = 1, 3y_1^2 + 2y_2^2 + 5y_3^2 + 2y_1y_2 - 2y_2y_3 = 1$.

1.3.20. Optimize $x_1y_1 + x_2y_1 + x_3y_1 + x_1y_2 + 3x_2y_2 + x_3y_2$ subject to $5x_1^2 + 2x_2^3 + 4x_3^2 - 2x_1x_2 - 4x_2x_3 = 1, 2y_1^2 + 3y_2^2 - 4y_1y_2 = 1$.

1.4. Derivative of a Scalar Function with Respect to a Matrix

Let $X = (x_{ij})$ be a $p \times p$ matrix with real and distinct elements. When x_{ij}'s are real and distinct we also call them functionally independent real variables. One can define all sorts of real-valued scalar functions of this X. For example, $\text{tr}(X) = $ trace of $X = x_{11} + \cdots + x_{pp}$ is a real-valued scalar function of X. $|X| = \det(X) = $ determinant of X is another real-valued scalar function of X. Let u be a real-valued scalar function of X. Then the derivative of u with respect to the matrix X is defined as

$$\frac{\partial u}{\partial X} = \left(\frac{\partial u}{\partial x_{ij}} \right),$$

that is, the (i, j)th element of $\frac{\partial u}{\partial X}$ is the partial derivative of u with respect to x_{ij}. For example, if $u = \text{tr}(X) = x_{11} + \cdots + x_{pp}$ then $\frac{\partial u}{\partial x_{ij}} = 0$ for $i \neq j$ and it is 1 when $i = j$. Hence we have the following result.

Result 1.4.1.

$$u = \text{tr}(X) \Rightarrow \frac{\partial u}{\partial X} = I$$

where I is the identity matrix.

Let

$$a = \begin{bmatrix} a_1 \\ \vdots \\ a_p \end{bmatrix}, \quad b = \begin{bmatrix} b_1 \\ \vdots \\ b_p \end{bmatrix}$$

be two $p \times 1$ vectors of constants. Consider the bilinear form $u = a'Xb$. For example, if $p = 2$ then

$$u = a'Xb = [a_1, a_2] \begin{bmatrix} x_{11} & x_{12} \\ x_{21} & x_{22} \end{bmatrix} \begin{bmatrix} b_1 \\ b_2 \end{bmatrix}$$

$$= a_1 x_{11} b_1 + a_1 x_{12} b_2 + a_2 x_{21} b_1 + a_2 x_{22} b_2,$$

then $\frac{\partial u}{\partial x_{11}} = a_1 b_1, \frac{\partial u}{\partial x_{12}} = a_1 b_2, \frac{\partial u}{\partial x_{21}} = a_2 b_1, \frac{\partial u}{\partial x_{22}} = a_2 b_2$ or

$$\frac{\partial u}{\partial X} = ab' = \begin{bmatrix} a_1 \\ \vdots \\ a_p \end{bmatrix} [b_1, \ldots, b_p]$$

and it is aa' when $b = a$ or when we have a quadratic form. In general, we can see that the term containing x_{ij} gives $a_i b_j$ and hence we have the following result.

Result 1.4.2. *Let $u = a'Xb$ where a and b are $p \times 1$, X is $p \times p$ of distinct real variables x_{ij}'s. Then*

$$\frac{\partial u}{\partial X} = ab'.$$

In this case, if $X = X'$ a symmetric matrix, then for $p = 2$ we have $u = a_1 x_{11} b_1 + a_1 x_{12} b_2 + a_2 x_{12} b_1 + a_2 x_{22} b_2$ and then $\frac{\partial u}{\partial x_{12}} = a_1 b_2 + a_2 b_1$ or $(1,2)$th element from $ab' + ba'$. But $\frac{\partial u}{\partial x_{ii}} = a_i b_i, i = 1, 2$ appearing only once. Then $\frac{\partial u}{\partial X} = ab' + ba' - \text{diag}(ab')$ when $X = X'$. This is our next result.

Result 1.4.3. *Let $u = a'Xb, a$ and b are $p \times 1$, X is $p \times p$, $X = X'$. Then*

$$u = a'Xb \Rightarrow \frac{\partial u}{\partial X} = ab' + ba' - \text{diag}(ab'),$$

where $\text{diag}(ab')$ means the diagonal matrix formed by the diagonal elements of ab'.

We can generalize the above result. Let A and B be $p \times p$ constant matrices and $X = (x_{ij})$ be a matrix of distinct x_{ij}'s, except for symmetry if X is symmetric. Then we have the following result.

Result 1.4.4. *Let A, B be $p \times p$ constant matrices and $X = (x_{ij})$ be a matrix of distinct real scalar variables, except for symmetry when $X = X'$. Let $u = \text{tr}(AX) = \text{tr}(XA)$. Then*

$$u = \text{tr}(AX) = \text{tr}(XA) \Rightarrow \frac{\partial u}{\partial X}$$

$$= \begin{cases} A' & \text{for a general } X, \\ A + A' - \text{diag}(A) & \text{for } X = X'. \end{cases}$$

$$v = \mathrm{tr}(AXB) \Rightarrow \frac{\partial v}{\partial X}$$

$$= \begin{cases} C' & \text{for a general } X, \\ C + C' - \mathrm{diag}(C) & \text{for } X = X', \end{cases}$$

where $C = BA$.

Result 1.4.5. *Let A be $q \times p$ and B be $p \times q$ constant matrices and let $X = (x_{ij})$ be $p \times p$ matrix of distinct real elements, except for symmetry if $X = X'$. Let $u = \mathrm{tr}(AXB)$. Then*

$$u = \mathrm{tr}(AXB) \Rightarrow \frac{\partial u}{\partial X}$$

$$= \begin{cases} A'B' & \text{for a general } X, \\ A'B' + BA - \mathrm{diag}(BA) & \text{if } X = X'. \end{cases}$$

Example 1.4.1. Let $a = \begin{bmatrix} 1 \\ 1 \end{bmatrix}$, $b = \begin{bmatrix} 1 \\ 0 \end{bmatrix}$, $A = \begin{bmatrix} 1 & 1 \\ 1 & 2 \end{bmatrix}$, $B = \begin{bmatrix} 1 & -1 \\ 1 & 1 \end{bmatrix}$, $X = \begin{bmatrix} x_{11} & x_{12} \\ x_{21} & x_{22} \end{bmatrix}$. Let $u_1 = a'Xb, u_2 = \mathrm{tr}(AX), u_3 = \mathrm{tr}(AXB)$. Compute $\frac{\partial u_j}{\partial X}, j = 1, 2, 3$ for a general X as well as when $X = X'$.

Solution 1.4.1.

$$u_1 = a'Xb = [1, 1] \begin{bmatrix} x_{11} & x_{12} \\ x_{21} & x_{22} \end{bmatrix} \begin{bmatrix} 1 \\ 0 \end{bmatrix} = x_{11} + x_{21}.$$

Then

$$\frac{\partial u_1}{\partial X} = \begin{bmatrix} 1 & 0 \\ 1 & 0 \end{bmatrix} = \begin{bmatrix} 1 \\ 1 \end{bmatrix} [1, 0] = ab'$$

and

$$\frac{\partial u_1}{\partial X} = \begin{bmatrix} 1 & 1 \\ 1 & 0 \end{bmatrix} \text{ for } X = X'.$$

But

$$ab' + ba' = \begin{bmatrix} 1 & 0 \\ 1 & 0 \end{bmatrix} + \begin{bmatrix} 1 & 1 \\ 0 & 0 \end{bmatrix} = \begin{bmatrix} 2 & 1 \\ 1 & 0 \end{bmatrix}$$

and diag$(ab') = $ diag$(1,0) \Rightarrow ab' + ba' - $ diag$(ab') = \begin{bmatrix} 1 & 1 \\ 1 & 0 \end{bmatrix}$. This verifies the result for $X = X'$. Now,

$$u_2 = \text{tr}\left\{\begin{bmatrix} 1 & 1 \\ 1 & 2 \end{bmatrix}\begin{bmatrix} x_{11} & x_{12} \\ x_{21} & x_{22} \end{bmatrix}\right\} = x_{11} + x_{21} + x_{12} + 2x_{22}.$$

Then

$$\frac{\partial u_2}{\partial X} = \begin{bmatrix} 1 & 1 \\ 1 & 2 \end{bmatrix} = A' \text{ for a general } X;$$

$$\frac{\partial u_2}{\partial X} = \begin{bmatrix} 1 & 2 \\ 2 & 2 \end{bmatrix} \text{ when } X = X'.$$

But

$$A + A' - \text{diag}(A) = \begin{bmatrix} 1 & 1 \\ 1 & 2 \end{bmatrix} + \begin{bmatrix} 1 & 1 \\ 1 & 2 \end{bmatrix} - \begin{bmatrix} 1 & 0 \\ 0 & 2 \end{bmatrix} = \begin{bmatrix} 1 & 2 \\ 2 & 2 \end{bmatrix}.$$

This verifies the result for $X = X'$. Consider u_3.

$$u_3 = \text{tr}(AXB) = \text{tr}(BAX) = \text{tr}\left\{\begin{bmatrix} 1 & -1 \\ 1 & 1 \end{bmatrix}\begin{bmatrix} 1 & 1 \\ 1 & 2 \end{bmatrix}X\right\}$$

$$= \text{tr}\left\{\begin{bmatrix} 0 & -1 \\ 2 & 3 \end{bmatrix}\begin{bmatrix} x_{11} & x_{12} \\ x_{21} & x_{22} \end{bmatrix}\right\} = -x_{21} + 2x_{12} + 3x_{22},$$

$$\frac{\partial u_3}{\partial X} = \begin{bmatrix} 0 & 2 \\ -1 & 3 \end{bmatrix} = A'B' = \begin{bmatrix} 1 & 1 \\ 1 & 2 \end{bmatrix}\begin{bmatrix} 1 & 1 \\ -1 & 1 \end{bmatrix}$$

$$= \begin{bmatrix} 0 & 2 \\ -1 & 3 \end{bmatrix} \text{ for a general } X.$$

When $X = X'$,

$$\frac{\partial u_3}{\partial X} = \begin{bmatrix} 0 & 1 \\ 1 & 3 \end{bmatrix}; \quad A'B' + BA = \begin{bmatrix} 0 & 2 \\ -1 & 3 \end{bmatrix} + \begin{bmatrix} 0 & -1 \\ 2 & 3 \end{bmatrix} = \begin{bmatrix} 0 & 1 \\ 1 & 6 \end{bmatrix}.$$

Hence

$$A'B' + BA - \text{diag}(BA) = \begin{bmatrix} 0 & 1 \\ 1 & 6 \end{bmatrix} - \begin{bmatrix} 0 & 0 \\ 0 & 3 \end{bmatrix} = \begin{bmatrix} 0 & 1 \\ 1 & 3 \end{bmatrix}.$$

This verifies the result for $X = X'$.

When X is non-singular, we can differentiate the determinant of X, that is, $|X|$, with respect to X. This is given as follows.

Result 1.4.6. *Let $X = (x_{ij})$ be a real $p \times p$ non-singular matrix of distinct real variables as elements. Let X^{-1} denote its regular inverse and $|X|$ its determinant. Then*

$$\frac{\partial |X|}{\partial X} = \begin{cases} |X|[X^{-1}]' & \text{for a general } X, \\ |X|[2X^{-1} - \text{diag}(X^{-1})] & \text{for } X = X'. \end{cases}$$

Proof. The determinant $|X|$ can be expanded as the elements and their cofactors of any row or column of X. That is,

$$|X| = x_{i1}|X_{i1}| + x_{i2}|X_{i2}| + \cdots + x_{ip}|X_{ip}|,$$

where $|X_{ij}|$ denotes the cofactor of x_{ij}. Then for a general X we have

$$\frac{\partial |X|}{\partial x_{ij}} = |X_{ij}| \Rightarrow \frac{\partial |X|}{\partial X} = \begin{bmatrix} |X_{11}| & \cdots & |X_{1p}| \\ \vdots & \cdots & \vdots \\ |X_{p1}| & \cdots & |X_{pp}| \end{bmatrix}.$$

This is the matrix of cofactors. But the inverse X^{-1} is the transpose of this cofactor matrix divided by the determinant. Hence

$$\frac{\partial |X|}{\partial X} = |X|[X^{-1}]'$$

for a general X. Note that when $X = X'$ we get $2|X_{ij}|$ for $i \neq j$ and one time $|X_{ii}|$ for $i = j$ when taking the partial derivatives. Hence we get the matrix

$$\frac{\partial |X|}{\partial X} = \begin{bmatrix} |X_{11}| & 2|X_{12}| & \cdots & 2|X_{1p}| \\ \vdots & \vdots & \cdots & \vdots \\ 2|X_{1p}| & 2|X_{2p}| & \cdots & |X_{pp}| \end{bmatrix}$$

$$= |X|[2X^{-1} - \text{diag}(X^{-1})], \quad \text{for } X = X'.$$

This completes the proof.

Example 1.4.2. Illustrate Results 1.4.6 for a 2×2 case.

Solution 1.4.2. Let $X = \begin{bmatrix} x_{11} & x_{12} \\ x_{21} & x_{22} \end{bmatrix}$, $|X| = x_{11}x_{22} - x_{12}x_{21} \neq 0$ for a general X and $|X| = x_{11}x_{22} - x_{12}^2 \neq 0$ for $X = X'$. Then

$$\frac{\partial |X|}{\partial X} = \begin{bmatrix} x_{22} & -x_{21} \\ -x_{12} & x_{11} \end{bmatrix}, \quad X^{-1} = \frac{1}{|X|}\begin{bmatrix} x_{22} & -x_{12} \\ -x_{21} & x_{11} \end{bmatrix}$$

or

$$|X|[[X^{-1}]' = \begin{bmatrix} x_{11} & -x_{21} \\ -x_{12} & x_{11} \end{bmatrix} = \frac{\partial |X|}{\partial X}.$$

This verifies the case for a general X. Now, consider the case $X = X'$. Here

$$\frac{\partial |X|}{\partial X} = \begin{bmatrix} x_{22} & -2x_{12} \\ -2x_{12} & x_{11} \end{bmatrix} = \begin{bmatrix} 2x_{22} & -2x_{12} \\ -2x_{12} & 2x_{11} \end{bmatrix} - \begin{bmatrix} x_{22} & 0 \\ 0 & x_{11} \end{bmatrix}$$

$$= |X|[2X^{-1} - \text{diag}(X^{-1})].$$

Thus the case, $X = X'$ is also verified.

Let us see the effect if the non-diagonal elements are multiplied by $\frac{1}{2}$. Let $Y = Y', y_{ij} = \frac{1}{2}x_{ij}, i \neq j, y_{ii} = x_{ii}, x_{ij} = x_{ji}$ for all i and j. Let us see what happens in a 2×2 case.

Example 1.4.3. Let $X^* = \begin{bmatrix} x_{11} & \frac{1}{2}x_{12} \\ \frac{1}{2}x_{12} & x_{22} \end{bmatrix}$. Compute $\frac{\partial |X^*|}{\partial X}$ when X is non-singular.

Solution 1.4.3. $|X^*| = x_{11}x_{22} - \frac{1}{4}x_{12}^2 \neq 0$.

$$\frac{\partial |X^*|}{\partial X^*} = \begin{bmatrix} \dfrac{\partial |X^*|}{\partial x_{11}} & \dfrac{\partial |X^*|}{\partial \left(\frac{1}{2}x_{12}\right)} \\[2ex] \dfrac{\partial |X^*|}{\partial \left(\frac{1}{2}x_{12}\right)} & \dfrac{\partial |X^*|}{\partial x_{22}} \end{bmatrix} = \begin{bmatrix} x_{22} & -x_{12} \\ -x_{12} & x_{11} \end{bmatrix} = |X|X^{-1}.$$

Result 1.4.7. *Let $X = (x_{ij}) = X'$. Let X^* be a $p \times p$ matrix of the type in Example 1.4.3 with non-diagonal elements weighted by $\frac{1}{2}$ and* diagonal elements as they are or weighted by 1. Then

$$\frac{\partial |X^*|}{\partial X^*} = |X| X^{-1}.$$

Example 1.4.4. Evaluate $\frac{\partial}{\partial X} \text{tr}(X^2)$ for a 2×2 case, both for a general X and for $X = X'$.

Solution 1.4.4.

$$\text{tr}(X^2) = \text{tr}\left\{ \begin{bmatrix} x_{11} & x_{12} \\ x_{21} & x_{22} \end{bmatrix} \begin{bmatrix} x_{11} & x_{12} \\ x_{21} & x_{22} \end{bmatrix} \right\}$$

$$= \begin{cases} x_{11}^2 + 2x_{12}x_{21} + x_{22}^2 & \text{for a general } X, \\ x_{11}^2 + 2x_{12}^2 + x_{22}^2 & \text{for } X = X'. \end{cases}$$

$$\frac{\partial \text{tr}(X^2)}{\partial X} = \begin{bmatrix} 2x_{11} & 2x_{21} \\ 2x_{12} & 2x_{22} \end{bmatrix} = 2X' \quad \text{for a general } X$$

and for $X = X'$,

$$\frac{\partial}{\partial X} \text{tr}(X^2) = \begin{bmatrix} 2x_{11} & 4x_{12} \\ 4x_{12} & 2x_{22} \end{bmatrix} = 4X - 2 \, \text{diag}(X)$$

$$= 2[2X - \text{diag}X] \quad \text{for } X = X'.$$

Hence we have the following result.

Result 1.4.8.

$$\frac{\partial}{\partial X} \text{tr}(X^2) = \begin{cases} 2X' & \text{for a general } X, \\ 4X - 2 \, \text{diag}(X) & \text{for } X = X'. \end{cases}$$

Result 1.4.9. *Let X be $p \times q$, A be $q \times q$ where A is a constant matrix and X is of distinct real variables. Then*

$$\frac{\partial}{\partial X} [\text{tr}(XAX')] = \begin{cases} X(A + A') & \text{for a general } X, \\ 2XA & \text{if } A = A'. \end{cases}$$

Proof. Let $x_{(1)}, x_{(2)}, \ldots, x_{(p)}$ be the rows of X. Then

$$XAX' = \begin{bmatrix} x_{(1)} \\ \vdots \\ x_{(p)} \end{bmatrix} A[x'_{(1)}, \ldots, x'_{(p)}] \Rightarrow \text{tr}(XAX')$$

$$= x_{(1)}Ax'_{(1)} + \cdots + x_{(p)}Ax'_{(p)}.$$

Then if we differentiate $\text{tr}(XAX')$ with the row $x_{(1)}$ then from Result 1.2.2 we have

$$\frac{\partial}{\partial x_{(1)}}\text{tr}(XAX') = x_{(1)}(A + A') \quad \text{or} \quad \frac{\partial}{\partial x_{(j)}} = x_{(j)}(A + A').$$

Now, stacking up the rows to get $\frac{\partial}{\partial X}\text{tr}(XAX') = X(A + A') = 2XA$ if $A = A'$.

Example 1.4.5. Verify Result 1.4.9 when $p = 2, q = 3$ and take a specific A.

Solution 1.4.5. Let

$$X = \begin{bmatrix} x_{11} & x_{12} & x_{13} \\ x_{21} & x_{22} & x_{23} \end{bmatrix}, \quad A = \begin{bmatrix} 1 & 1 & 1 \\ 1 & -1 & 1 \\ -1 & 1 & -1 \end{bmatrix}.$$

Then

$$\text{tr}(XAX') = \text{tr}\left\{ \begin{bmatrix} x_{11} & x_{12} & x_{13} \\ x_{21} & x_{22} & x_{23} \end{bmatrix} \begin{bmatrix} 1 & 1 & 1 \\ 1 & -1 & 1 \\ -1 & 1 & -1 \end{bmatrix} \begin{bmatrix} x_{11} & x_{21} \\ x_{12} & x_{22} \\ x_{13} & x_{23} \end{bmatrix} \right\}$$

$$= (x_{11} + x_{12} + x_{13})x_{11} + (x_{11} - x_{12} + x_{13})x_{12}$$

$$+ (x_{11} + x_{12} - x_{13})x_{13}$$

$$+ (x_{21} + x_{22} + x_{23})x_{21} + (x_{21} - x_{22} + x_{23})x_{22}$$

$$+ (x_{21} + x_{22} - x_{23})x_{23},$$

$$\frac{\partial}{\partial X}\text{tr}(XAX') = \begin{bmatrix} (x_{11}, x_{12}, x_{13}) \\ (x_{21}, x_{22}, x_{23}) \end{bmatrix} \begin{bmatrix} 2 & 2 & 0 \\ 2 & -2 & 2 \\ 0 & 2 & -2 \end{bmatrix} = X(A + A').$$

Note that if we have a one-to-one function of $|X|$, say $\phi(|X|)$ then when we differentiate with respect to a scalar x_{ij} we have $\frac{\partial}{\partial x_{ij}}\phi(|X|) = \phi'(|X|)\frac{\partial}{\partial x_{ij}}|X|$. Hence, in general, we have the following result.

Result 1.4.10.

$$\frac{\partial}{\partial X}\phi(|X|) = \phi'(|X|)\begin{cases} |X|(X^{-1})' & \text{for a general } X \\ |X|[2X^{-1} - \text{diag}(X^{-1})] & \text{for } X = X'. \end{cases}$$

Some problems of this type are given in the exercises.

Exercise 1.4

1.4.1. Let $X = (x_{ij})$, $p \times p$. Let $a' = (1,1,1), b' = (1,-1,1), p = 3$. Compute $\frac{\partial}{\partial X}(a'Xb)$ for (i) general X, (ii) for $X = X'$.

1.4.2. Let $X = (x_{ij})$ be 4×4, $a' = (1,1,-1,1), b' = (1,-1,1,-1)$. Compute $\frac{\partial}{\partial X}(a'Xb)$ for (i) general X, (ii) for $X = X'$.

1.4.3. Let $X = (x_{ij})$ be 3×3. Let

$$A_1 = \begin{bmatrix} 1 & 2 & 1 \\ 1 & -1 & 1 \\ 2 & 1 & 3 \end{bmatrix}, \quad A_2 = \begin{bmatrix} 1 & 1 & 1 \\ 2 & -1 & 1 \\ 1 & -3 & 2 \end{bmatrix}, \quad A_3 = \begin{bmatrix} 1 & 1 & 1 \\ 1 & 2 & -2 \\ 1 & -2 & 5 \end{bmatrix}.$$

Compute $\frac{\partial}{\partial X}\text{tr}(A_j X) = \frac{\partial}{\partial X}\text{tr}(X A_j), j = 1, 2, 3$ for the cases: (i) X is general, (ii) $X = X'$.

1.4.4. Let $X = (x_{ij})$ and 2×2. Let

$$A_1 = \begin{bmatrix} 2 & -2 \\ 1 & 5 \end{bmatrix}, \quad B_1 = \begin{bmatrix} 1 & 1 \\ 0 & 1 \end{bmatrix},$$

$$A_2 = \begin{bmatrix} 1 & -1 \\ -1 & 1 \end{bmatrix}, \quad B_2 = \begin{bmatrix} 1 & 2 \\ 2 & 4 \end{bmatrix}.$$

Compute $\frac{\partial}{\partial X}\text{tr}(A_j X B_j), j = 1, 2$ and for the cases: (i) general X, (ii) for $X = X'$.

1.4.5. Let $X = (x_{ij})$ be 3×3. Let

$$A_1 = \begin{bmatrix} 1 & -1 & 1 \\ 1 & 2 & 1 \\ -1 & 1 & 1 \end{bmatrix}, \quad B_1 = \begin{bmatrix} 2 & 0 & -1 \\ 0 & 4 & 1 \\ -1 & 1 & 3 \end{bmatrix},$$

$$A_2 = \begin{bmatrix} 1 & -1 & 1 \\ -1 & 2 & 3 \\ 1 & 3 & 4 \end{bmatrix}, \quad B_2 = \begin{bmatrix} 2 & 1 & -1 \\ 1 & 4 & 3 \\ -1 & 3 & 2 \end{bmatrix}.$$

Compute $\frac{\partial}{\partial X}\operatorname{tr}(A_j X B_j), j = 1,2$ for the cases: (i) general X, (ii) for $X = X'$.

1.4.6. Let (a) $X = (x_{ij}), 2 \times 2$ and (b) $X = (x_{ij})$, 3×3. Let X be non-singular in each case. Compute $\frac{\partial}{\partial X}|X|$ for each case for the situations (i) X is general, (ii) $X = X'$.

1.4.7. Let (a) X^* be 2×2 and (b) X^* be 3×3 and non-singular in each case, where the off-diagonal elements are weighted by $\frac{1}{2}$. Compute $\frac{\partial}{\partial X^*}|X^*|$ in each case for the situations (i) X is general, (ii) $X = X'$.

1.4.8. Let (a) $u = \mathrm{e}^{-2|X|}$, (b) $u = \mathrm{e}^{2|X|+5}$. Compute $\frac{\partial u}{\partial X}$ for each case and for the situations (i) X is general, (ii) $X = X'$.

1.4.9. Let (a) $u = \ln|X|, |X| > 0$, (b) $u = 3\ln|X| + 5|X|^2, |X| > 0$. Compute $\frac{\partial u}{\partial X}$ in each case and for situations where (i) X is general, (ii) $X = X'$.

1.4.10. Let (a) $u = \mathrm{e}^{-\operatorname{tr}(X^2)}$, (b) $u = \mathrm{e}^{-\operatorname{tr}(X^2)-3|X|}$. Compute $\frac{\partial u}{\partial X}$ in each case and for situations (i) X is general, (ii) $X = X'$.

1.4.11. Maximum likelihood estimates. Let X be a $p \times 1$ vector of real variables so that $X' = (x_1, \ldots, x_p)$. Then X is said to have a multivariate Gaussian density if the density function of X (*Note:* A density is a non-negative integrable function with total integral unity), is of the form

$$f(X) = \frac{\mathrm{e}^{-\frac{1}{2}(X-\mu)'V^{-1}(X-\mu)}}{(2\pi)^{\frac{p}{2}}|V|^{\frac{1}{2}}}, \quad V = V' > O,$$

is a real positive definite constant matrix, μ is a $p \times 1$ constant vector of unknown parameters, V is also unknown. Consider observations on X, the jth observation vector be denoted by $X_j, X_j' = (x_{1j}, x_{2j}, \ldots, x_{pj}), j = 1, \ldots, n$. Then the likelihood function L is

$$L = \prod_{j=1}^{n} \frac{e^{-\frac{1}{2}(X_j - \mu)'V^{-1}(X_j - \mu)}}{(2\pi)^{\frac{p}{2}}|V|^{\frac{1}{2}}} = \frac{e^{-\frac{1}{2}\sum_{j=1}^{n}(X_j - \mu)'V^{-1}(X_j - \mu)}}{(2\pi)^{\frac{np}{2}}|V|^{\frac{n}{2}}}.$$

Maximizing L also means maximizing $\ln L$. Maximize L and obtain an estimate $\hat{\mu}$ for μ when (i) V is known, (ii) obtain estimates of μ and V, (iii) obtain the maximum value of L.

Hint: If you are using calculus then solve for μ and V from the equations $\frac{\partial \ln L}{\partial \mu} = O, \frac{\partial \ln L}{\partial V} = O$ where O indicates a null vector/matrix, respectively. This is the case when both μ and V are unknown.

1.4.12. Let

$$f(X) = \frac{|B|^{\alpha}}{\Gamma_p(\alpha)}|X|^{\alpha - \frac{p+1}{2}}e^{-(BX)}, X = X' > O, B = B' > O,$$

and B and the scalar α are unknown parameters. This $f(X)$ is known as a real matrix-variate gamma density where X and B are $p \times p$ and positive definite, B is constant matrix. Let the jth observation on the $p \times p$ matrix X be denoted by $X_j, j = 1, \ldots, n$. Then the likelihood function L is obtained as

$$L = \prod_{j=1}^{n} f(X_j).$$

Maximize L or $\ln L$ and obtain an estimate for B when α is known, where $\Gamma_p(\alpha)$ is the real matrix-variate gamma, given by

$$\Gamma_p(\alpha) = \pi^{\frac{p(p-1)}{4}}\Gamma(\alpha)\Gamma\left(\alpha - \frac{1}{2}\right)\cdots\Gamma\left(\alpha - \frac{p-1}{2}\right), \quad \Re(\alpha) > \frac{p-1}{2}.$$

Acknowledgments

The author would like to thank the Department of Science and Technology, Government of India, for the financial assistance for this work under project number SR/S4/MS:287/05 and the Centre for Mathematical and Statistical Sciences for the facilities.

Bibliography

[1] A.M. Mathai, *Jacobians of Matrix Transformations and Functions of Matrix Argument*, World Scientific Publishing, New York, 1997.

[2] A.M. Mathai, A pathway to matrix-variate gamma and normal densities, *Linear Algebra Appl.* **396** (2005), 317–328.

[3] A.M. Mathai, Some properties of Mittag-Leffler functions and matrix-variate analogues: A statistical perspective, *Fract. Calc. Appl. Anal.* **13**(1) (2010), 113–132.

[4] A.M. Mathai and H.J. Haubold, Kober operators from a statistical perspective I–IV, preprint, Cornell arXiv.

[5] A.M. Mathai and H.J. Haubold, *Special Functions for Applied Scientists*, Springer, New York, 2008.

[6] A.M. Mathai and H.J. Haubold, Fractional operators in the matrix variate case, *Fract. Cal. Appl. Anal.* **16**(2) (2013), 469–478.

[7] A.M. Mathai, R.K. Saxena and H.J. Haubold, *The H-function: Theory and Applications*, Springer, New York, 2010.

[8] S. Kounlas and A.M Mathai, Maximizing the sum of integers when their sum of squares in fixed, *Optimization* **19**(1988), 123–131.

Chapter 2

Jacobians of Matrix Transformations
and Functions of Matrix Argument*

2.1. Introduction

In Chapter 1 we have defined wedge product of differentials and Jacobians. When $y_j = f_j(x_1, \ldots, x_k)$ where $x_1, \ldots, x_k$ are real scalar variables and $f_j, j = 1, \ldots, k$ are real-valued scalar functions of $x_1, \ldots, x_k$ then the connection between the wedge product of differentials, $dx_1, \ldots, dx_k$ and $dy_1, \ldots, dy_k$, is given by

$$dY = dy_1 \wedge \cdots \wedge dy_k = J \, dx_1 \wedge \cdots \wedge dx_k = J \, dX,$$

where J is the determinant

$$J = \left| \left(\frac{\partial y_i}{\partial x_j} \right) \right|$$

with $\left(\frac{\partial y_i}{\partial x_j} \right)$ denoting the matrix of partial derivative of y_i with respect to $x_j, i, j = 1, \ldots, k$. If this determinant J is non-zero then $dY = J \, dX \Rightarrow dX = \frac{1}{J} dY$. These basic ideas can be generalized for transformations involving matrices. We will use the general notation dX when X is scalar, vector or matrix. If $X = (x_{ij})$ is a $p \times q$ matrix then

$$dX = \bigwedge_{i=1}^{p} \bigwedge_{j=1}^{q} dx_{ij} \qquad (2.1.1)$$

---

*This chapter is summarized from the lectures given by Professor Dr A.M. Mathai.

the elements can be taken in any given order but that order has to be maintained until the computations are over. For each transposition of two differentials the resulting quantity has to be multiplied by -1. First, we will consider linear transformations.

2.2. Linear Transformations

First, we consider the simplest of the linear transformations. Let

$$y_1 = a_{11}x_1 + a_{12}x_2 + \cdots + a_{1k}x_k$$

$$y_2 = a_{21}x_1 + a_{22}x_2 + \cdots + a_{2k}x_k$$

$$\vdots$$

$$y_k = a_{k1}x_1 + a_{k2}x_2 + \cdots + a_{kk}x_k$$

or

$$Y = AX, X = \begin{bmatrix} x_1 \\ \vdots \\ x_k \end{bmatrix}, \quad Y = \begin{bmatrix} y_1 \\ \vdots \\ y_k \end{bmatrix}, \quad A = \begin{bmatrix} a_{11} & \cdots & a_{1k} \\ \vdots & \cdots & \vdots \\ a_{k1} & \cdots & a_{kk} \end{bmatrix},$$

where the a_{ij}'s are constants and x_j's are real scalar variables. If A is non-singular then the regular inverse A^{-1} exists, then $Y = AX \Rightarrow X = A^{-1}Y$ or the transformation is one-to-one. It is easy to see that $\frac{\partial y_i}{\partial x_j} = a_{ij}$ and hence $dY = |A| \, dX$. We may also take the differentials of y_j, $j = 1, \ldots, k$ and then take the wedge product to get the same result. We may also write the matrix of partial derivatives as

$$\left(\frac{\partial y_i}{\partial x_j} \right) = \frac{\partial Y}{\partial X} = A \qquad (2.2.1)$$

in this case.

Result 2.2.1. *Let X and Y be $p \times 1$ vectors of distinct real variables and let $A = (a_{ij})$ be a $p \times p$ non-singular matrix of constants. Then*

$$Y = AX, |A| \neq 0 \Rightarrow dY = |A|dX. \qquad (2.2.2)$$

Example 2.2.1. Evaluate the Jacobian in the transformation $y_1 = 2x_1 - x_2 + x_3$, $y_2 = x_1 + x_2 + x_3$, $y_3 = 3x_1 + x_2 - x_3$.

Solution 2.2.1. Writing the linear equations together as $Y = AX$ we have

$$X = \begin{bmatrix} x_1 \\ x_2 \\ x_3 \end{bmatrix}, \quad Y = \begin{bmatrix} y_1 \\ y_2 \\ y_3 \end{bmatrix}, \quad A = \begin{bmatrix} 2 & -1 & 1 \\ 1 & 1 & 1 \\ 3 & 1 & -1 \end{bmatrix}, \quad |A| = -10 \neq 0.$$

$$\frac{\partial y_1}{\partial x_1} = 2, \quad \frac{\partial y_1}{\partial x_2} = -1, \quad \frac{\partial y_1}{\partial x_3} = 1, \ldots \text{ or } dY = |A|\, dX = -10\, dX.$$

Example 2.2.2. Evaluate the integral

$$u = \int_X e^{-X'A'AX} dX,$$

where the integral is over all X, $-\infty < x_j < \infty$, $j = 1, 2, 3$ where X and A are as defined in Example 2.2.1.

Solution 2.2.2. Making the transformation $Y = AX \Rightarrow dY = |A|\, dX$ or $dX = \frac{1}{|A|} dY$. Also, $X'A'AX = Y'Y = y_1^2 + y_2^2 + y_3^2$. Since the exponent is symmetric, free of the order in which y_j's are taken and since the exponential integral is positive we replace dX by $\frac{1}{10} dY$. But $\int_{-\infty}^{\infty} e^{-y_j^2} dy_j = \sqrt{\pi}$, then we have the final result as

$$u = \frac{\pi^{\frac{3}{2}}}{10}.$$

A more general linear transformation can be given in terms of an $m \times n$ matrix X going to an $m \times n$ matrix Y. Consider the transformation $Y = AX$, $|A| \neq 0$ where X and Y are $m \times n$, A is $m \times m$ and non-singular. Then we have a one-to-one transformation and have the following result.

Result 2.2.2. *Let X and Y be $m \times n$ matrices of distinct real scalar variables and let A be $m \times m$ non-singular constant matrix. Then*

$$Y = AX \Rightarrow X = A^{-1}Y \quad \text{and} \quad dY = |A|^n dX. \tag{2.2.3}$$

Proof. Consider the columns of X. Let the columns of X and Y be denoted by $X^{(j)}, Y^{(j)}, j = 1, \ldots, n$, respectively. Then

$$Y = (Y^{(1)}, \ldots, Y^{(n)}) = AX = A(X^{(1)}, \ldots, AX^{(n)}).$$

That is, $Y^{(j)} = AX^{(j)}, j = 1, \ldots, n$. Note that $\frac{\partial Y^{(i)}}{\partial X^{(i)}} = A$ and $\frac{\partial Y^{(i)}}{\partial X^{(j)}} = O, i \neq j$. Now, consider the chain of variables $\tilde{X} = \begin{bmatrix} X^{(1)} \\ \vdots \\ X^{(n)} \end{bmatrix}$

and the corresponding $\tilde{Y} = \begin{bmatrix} Y^{(1)} \\ \vdots \\ Y^{(n)} \end{bmatrix}$. Then the Jacobian matrix of this transformation is

$$\frac{\partial \tilde{Y}}{\partial \tilde{X}} = \begin{bmatrix} A & O & \cdots & O \\ \vdots & \vdots & \cdots & \vdots \\ O & O & \cdots & A \end{bmatrix}.$$

It is a block diagonal matrix with n diagonal blocks, each having the matrix A and hence the determinant is $|A|^n$. This establishes the result.

Suppose that X is postmultiplied by an $n \times n$ non-singular matrix B then we have the following result.

Result 2.2.3. *Let X and Y be $m \times n$ matrices of distinct real scalar variables and let B be a $n \times n$ non-singular constant matrix. Then*

$$Y = XB \Rightarrow dY = |B|^m dX. \qquad (2.2.4)$$

Proof. Here we consider the rows of X and Y. Let the rows be denoted by

$$X = \begin{bmatrix} X_{(1)} \\ \vdots \\ X_{(m)} \end{bmatrix}, \quad Y = \begin{bmatrix} Y_{(1)} \\ \vdots \\ Y_{(m)} \end{bmatrix},$$

$$Y = XB \Rightarrow Y_{(i)} = X_{(i)}B, i = 1, \ldots, m.$$

Then the Jacobian matrix is a block diagonal matrix with m diagonal blocks equal to B each and thus the Jacobian is $|B|^m$.

Now, combining the above two results we have the most general linear transformation and the following result.

Result 2.2.4. *Let X and Y be $m \times n$ matrices of distinct real scalar variables. Let A and B be $m \times m$ and $n \times n$ non-singular constant matrices, respectively. Then*

$$Y = AXB \Rightarrow dY = |A|^n |B|^m dX. \qquad (2.2.5)$$

The proof is trivial. Let $Y = AXB = AU, U = XB$. Now, apply Results 2.2.2 and 2.2.3 to establish Result 2.2.4.

Example 2.2.3. Let

$$X = \begin{bmatrix} x_{11} & x_{12} & x_{13} \\ x_{21} & x_{22} & x_{23} \end{bmatrix}, \quad A = \begin{bmatrix} 1 & 1 \\ 1 & 2 \end{bmatrix}, \quad B = \begin{bmatrix} 1 & 1 & 1 \\ 2 & -1 & 1 \\ -1 & 2 & -1 \end{bmatrix}.$$

Consider the transformations $Y_1 = AX, Y_2 = XB, Y_3 = AXB$. Write dY_1, dY_2, dY_3 in terms of dX.

Solution 2.2.3. Let the columns of X and Y be $X^{(j)}, Y^{(j)}, j = 1, 2, 3$ respectively. Then $Y^{(j)} = AX^{(j)}, j = 1, 2, 3$. Also, then

$$\frac{\partial Y^{(i)}}{\partial X^{(j)}} = \begin{cases} O, & i \neq j, \\ A, & i = j. \end{cases}$$

The Jacobian matrix is

$$\begin{bmatrix} A & O & O \\ O & A & O \\ O & O & A \end{bmatrix}, \quad J = |A|^3 = 1^3 = 1.$$

Therefore $dY_1 = dX$. Let the rows of X and Y be denoted by

$$X = \begin{bmatrix} X_{(1)} \\ X_{(2)} \end{bmatrix}, \quad Y = \begin{bmatrix} Y_{(1)} \\ Y_{(2)} \end{bmatrix}, \quad Y_2 = XB \Rightarrow Y_{(j)} = X_{(j)}B, \quad j = 1, 2.$$

Then

$$\frac{\partial Y'_{(i)}}{\partial X'_{(j)}} = \begin{cases} B, & i = j, \\ O, & i \neq j. \end{cases}$$

The Jacobian matrix is

$$\begin{bmatrix} B & O \\ O & B \end{bmatrix} \Rightarrow J = |B|^2 = 3^2 = 9.$$

Hence $dY_2 = 9\,dX$. Now, consider $Y_3 = AXB = AU, U = XB$. Then $dY_3 = |A|^3 dU, dU = |B|^2 dX \Rightarrow dY_3 = |A|^3|B|^2 dX = (1)^3(3)^2 dX = 9\,dX$.

When $m = n = p$ we have $Y = AXB \Rightarrow dY = |A|^p|B|^p dX$. What happens if $B = A'$ and $X = X'$ so that $Y = Y'$? In other words, what will be the Jacobian if the transformation is linear but involving symmetric matrices. Obviously, we cannot use the earlier results because now we have only $\frac{p(p+1)}{2}$ real variables in a $p \times p$ symmetric matrix instead of p^2 distinct elements in a $p \times p$ matrix of distinct elements. Transformations involving symmetric matrices are the most commonly occurring linear transformations. We have the following result.

Result 2.2.5. *Let* $Y = AXA', |A| \neq 0$. *Let* X *and* Y *be* $p \times p$ *symmetric matrices of distinct real variables except for symmetry, or with* $\frac{p(p+1)}{2}$ *distinct variables each. Then*

$$Y = AXA', |A| \neq 0 \Rightarrow dY$$

$$= \begin{cases} |A|^{p+1} dX & \text{for } X = X' \text{ symmetric,} \\ |A|^{p-1} dX & \text{for } X = -X' \text{ skew symmetric.} \end{cases} \tag{2.2.6}$$

Proof. The proof is not straightforward. It cannot be proved with the help of the earlier results. We will use the property that a non-singular matrix can be written as a product of basic elementary matrices. Let $A = E_1 E_2 F_1 \ldots E_k F_r$ where the E's and F's are the basic elementary matrices. Then $Y = AXA' = E_1E_2F_1 \ldots E_kF_rXF_r'$ $E_k' \ldots E_1'$. An E-type elementary matrix is obtained by multiplying any row (column) of an identity matrix by a non-zero scalar and an F-type elementary matrix is obtained by adding any row (column) to any other row (column) of an identity matrix. For example, E-type and F-type 3×3 elementary matrices are as follows:

$$E_j = \begin{bmatrix} 1 & 0 & 0 \\ 0 & 5 & 0 \\ 0 & 0 & 1 \end{bmatrix}, \quad F_i = \begin{bmatrix} 1 & 0 & 0 \\ 1 & 1 & 0 \\ 0 & 0 & 1 \end{bmatrix}$$

for some j and i, where E_j is obtained by multiplying the second row of an identity matrix by the non-zero scalar, namely 5, and F_i is obtained by 'adding the first row to the second row of an identity matrix. Let $Y_1 = F_rXF_r', Y_2 = E_kY_1E_k', \ldots$. Then dY_1 is evaluated in terms of dX, and then dY_2 in terms of dY_1 which is available in terms of dX etc. and finally dY in terms of dX. An E-type transformation; for example, $Y_t = E_jY_{t-1}E_j'$, for some Y_t will produce the following Jacobian. Note that the second row and second column of Y_{t-1} are multiplied by 5. When taking the wedge product, we are only counting the diagonal elements and either the elements on the right of the diagonal or on the left of the diagonal. One diagonal element is multiplied twice. In the case of the above E_j the second diagonal element is multiplied by 5^2 and other elements by only once. Thus, 5 comes $3 + 1 = 4$ times when taking the wedge product. In the above case, $dY_t = 5^4dY_{t-1}$. In general if the ith row of a $p \times p$ identity matrix is multiplied by $c \neq 0$ then the corresponding Jacobian gives $c^{p+1} = |E|^{p+1}$. In the case of a F-type matrix the Jacobian will be $1 = 1^{p+1} = |F|^{p+1}$. Taking all the products, the Jacobian $J = |E_1|^{p+1} \times |E_2|^{p+1} \times \cdots \times |F_r|^{p+1} = |A|^{p+1}$ when $X = X'$. When X is skew symmetric, the diagonal elements are zeros and thus the exponent will be $(p+1) - 2 = p - 1$.

Example 2.2.4. Evaluate the integral $u = \int_X e^{-\text{tr}(AXBX')}dX$ where X is $p \times q$, A is $p \times p$ and B is $q \times q$ positive definite constant matrices and X is a matrix of pq distinct real scalar variables.

Solution 2.2.4. Since A and B are positive definite their unique symmetric square roots exist, denoted by $A^{\frac{1}{2}}$ and $B^{\frac{1}{2}}$, respectively. Note that for any two matrices G and H where GH and HG are defined, $\operatorname{tr}(GH) = \operatorname{tr}(HG)$. Hence

$$\operatorname{tr}(AXBX') = \operatorname{tr}(A^{\frac{1}{2}}XB^{\frac{1}{2}}B^{\frac{1}{2}}X'A^{\frac{1}{2}}) = \operatorname{tr}(YY'), Y = A^{\frac{1}{2}}XB^{\frac{1}{2}}.$$

From Result 2.2.4 we have $\mathrm{d}Y = |A|^{\frac{q}{2}}|B|^{\frac{p}{2}}\mathrm{d}X$. But for any matrix G, $\operatorname{tr}(GG') = $ the sum of squares of all elements in G. Therefore $\operatorname{tr}(YY') = \sum_{i=1}^{p}\sum_{j=1}^{q} y_{ij}^2$. Also $\int_{-\infty}^{\infty} e^{-y_{ij}^2}\mathrm{d}y_{ij} = \sqrt{\pi}$. Hence

$$u = \frac{\pi^{\frac{pq}{2}}}{|A|^{\frac{q}{2}}|B|^{\frac{p}{2}}}$$

and

$$f(X) = \frac{|A|^{\frac{q}{2}}|B|^{\frac{p}{2}}}{\pi^{\frac{pq}{2}}}e^{-\operatorname{tr}(AXBX')}, \quad A > O, \quad B > O, \quad X = (x_{ij}), \ p \times q$$

is known as the real matrix-variate Gaussian density, so that, $f(X) \geq 0$ for all X and $\int_X f(X)\mathrm{d}X = 1$. Note that we can also replace X by $X - M$ where M is a $p \times q$ constant matrix. Moreover, $\mathrm{d}(X - M) = \mathrm{d}X$.

Note that if C is a constant $m \times n$ matrix, A is an $m \times m$ non-singular constant matrix, a is a real scalar then

$$Y = X + C \Rightarrow \mathrm{d}Y = \mathrm{d}X,$$

$$Y = AX + C \Rightarrow \mathrm{d}Y = |A|^n\mathrm{d}X,$$

$$Y = aX + C \Rightarrow \mathrm{d}Y = \begin{cases} a^{mn}\mathrm{d}X & \text{when } X \text{ is general,} \\ a^{\frac{p(p+1)}{2}}\mathrm{d}X & \text{when } X = X' \text{ and } p \times p. \end{cases}$$

$$(2.2.7)$$

Some interesting Jacobians can be obtained when X and A are lower triangular matrices.

Result 2.2.6. *Let X and A be lower triangular $p \times p$ matrices, $X = (x_{ij}), i \geq j$, be of distinct real scalar variables and let $A = (a_{ij}), i \geq j$, be a constant matrix with $a_{jj} > 0, j = 1, \ldots, p$. Then*

$$Y = X + X' \Rightarrow \mathrm{d}Y = 2^p \mathrm{d}X,$$

$$Y = XA \Rightarrow \mathrm{d}Y = \left\{ \prod_{j=1}^{p} a_{jj}^{p-j+1} \right\} \mathrm{d}X,$$

$$Y = AX \Rightarrow \mathrm{d}Y = \left\{ \prod_{j=1}^{p} a_{jj}^{j} \right\} \mathrm{d}X,$$

$$Y = aX \Rightarrow \mathrm{d}Y = a^{\frac{p(p+1)}{2}} \mathrm{d}X \text{ where } a \text{ is a real scalar.}$$

Proof. In $X + X'$ the diagonal elements are multiplied by 2 each and all other elements are multiplied by 1 each. The distinct elements are $x_{ij}, y_{ij}, \ i \geq j$ only. When taking the wedge product of differentials, $\mathrm{d}y_{jj} = 2\mathrm{d}x_{jj}$ and $\mathrm{d}y_{ij} = \mathrm{d}x_{ij}, i > j$. Then $\mathrm{d}Y = \bigwedge_{i \geq j} \mathrm{d}y_{ij} = 2^p \bigwedge_{i \geq j} \mathrm{d}x_{ij}$. Note that

$$Y = XA$$

$$= \begin{bmatrix} x_{11} & 0 & \cdots & 0 \\ x_{21} & x_{22} & \cdots & 0 \\ \vdots & \vdots & \cdots & \vdots \\ x_{p1} & x_{p2} & \cdots & x_{pp} \end{bmatrix} \begin{bmatrix} a_{11} & 0 & \cdots & 0 \\ a_{21} & a_{22} & \cdots & 0 \\ \vdots & \vdots & \cdots & \vdots \\ a_{p1} & a_{p2} & \cdots & a_{pp} \end{bmatrix}$$

$$= \begin{bmatrix} a_{11}x_{11} & 0 & \cdots & 0 \\ a_{11}x_{21} + a_{21}x_{22} & a_{22}x_{22} & \cdots & 0 \\ \vdots & \vdots & \cdots & \vdots \\ \sum_{k=1}^{p} x_{pk}a_{k1} & \sum_{k=2}^{p} x_{pk}a_{k2} & \cdots & x_{pp}a_{pp} \end{bmatrix}.$$

Take y_{ij}'s in the order $y_{11}, (y_{21}, y_{22}), \ldots, (y_{p1}, \ldots, y_{pp})$ and x_{ij}'s in the same order. Then

$$\frac{\partial y_{11}}{\partial x_{11}} = a_{11}, \quad \frac{\partial (y_{21}, y_{22})}{\partial (x_{21}, x_{22})} = \begin{bmatrix} a_{11} & a_{21} \\ 0 & a_{22} \end{bmatrix}, \ldots, \frac{\partial (y_{p1}, \ldots, y_{pp})}{\partial (x_{p1}, \ldots, x_{pp})}$$

$$= \begin{bmatrix} a_{11} & a_{21} & \cdots & a_{p1} \\ 0 & a_{22} & \cdots & a_{p2} \\ \vdots & \vdots & \cdots & \vdots \\ 0 & 0 & \cdots & a_{pp} \end{bmatrix}.$$

Taking the determinants and then products of determinants we have

$$dY = a_{11}^p a_{22}^{p-1} \cdots a_{pp} dX = \left\{ \prod_{j=1}^{p} a_{jj}^{p+1-j} \right\} dX.$$

When $Y = AX$ note that the roles of a_{ij}'s and x_{ij}'s are interchanged in the explicit form given for XA above. Then the matrices of partial derivatives will be triangular with diagonal elements a_{11} appearing once, a_{22} repeated twice and so on. Hence the Jacobian is $\prod_{j=1}^{p} a_{jj}^j$. Since a lower triangular matrix has only $\frac{p(p+1)}{2}$ distinct non-zero elements, each multiplied by a scalar quantity a gives the Jacobian $a^{\frac{p(p+1)}{2}}$ since the matrix is $p \times p$.

By taking the transposes and observing that $dY = dY'$, results for upper triangular matrices can be written down from Result 2.2.6. This is given in the following exercises.

Exercise 2.2

2.2.1. Let $Y = aX$ where Y and X are $p \times 1$ and a is a scalar. Show that $dY = a^p dX, dY' = a^p dX'$.

2.2.2. Verify Result 2.2.1 for the equation

$$Y = \begin{bmatrix} y_1 \\ y_2 \\ y_3 \end{bmatrix} = \begin{bmatrix} 2 & -3 & -1 \\ -3 & 4 & 0 \\ -1 & 0 & 8 \end{bmatrix} \begin{bmatrix} x_1 \\ x_2 \\ x_3 \end{bmatrix} = AX.$$

2.2.3. Verify Results 2.2.2–2.2.4 for $A = \begin{bmatrix} 2 & 1 \\ 1 & 3 \end{bmatrix}$, $B = \begin{bmatrix} 3 & -1 \\ -1 & 2 \end{bmatrix}$.

2.2.4. Verify Results 2.2.3–2.2.4 for

$$A = \begin{bmatrix} 3 & 0 & 1 \\ 0 & 2 & -1 \\ 1 & -1 & 5 \end{bmatrix}, \quad B = \begin{bmatrix} 1 & 1 & 1 \\ 1 & 2 & 2 \\ 1 & 2 & 4 \end{bmatrix}.$$

2.2.5. Verify Result 2.2.5 for

$$(a): \begin{bmatrix} 1 & 1 & 1 \\ 1 & 2 & -1 \\ -2 & 0 & 4 \end{bmatrix}, \quad (b): A = \begin{bmatrix} 1 & 0 & 1 \\ 0 & 2 & -1 \\ 1 & -1 & 4 \end{bmatrix}$$

when (1): $X = X'$, (2): $X = -X'$.

2.2.6. Show that when X is skew symmetric and $p \times p$, $Y = aX \Rightarrow dY = a^{\frac{p(p-1)}{2}} dX$ where a is a real scalar constant.

2.2.7. Let X and A be lower triangular $p \times p$ matrices, $X = (x_{ij})$, $i \geq j$, is a matrix of distinct real scalar variables and A is a non-singular constant matrix with positive diagonal elements. Then show that

$$Y = A'X + X'A \Rightarrow dY = \left\{ \prod_{j=1}^{p} a_{jj}^{p} \right\} dX,$$

$$Y = AX' + XA' \Rightarrow dY = 2^p \left\{ \prod_{j=1}^{p} a_{jj}^{p-j+1} \right\} dX,$$

$$Y = XA^{-1} + (A')^{-1}X' \Rightarrow dY = 2^p \left\{ \prod_{j=1}^{p} a_{jj}^{-(p-j+1)} \right\} dX.$$

2.2.8. Let X, A be $p \times p$ upper triangular matrices with $a_{jj} > 0$, $j = 1, \ldots, p$ and x_{ij}'s are distinct real scalar variables. Then show that

$$Y = X + X' \Rightarrow dY = 2^p dX,$$

$$Y = XA \Rightarrow dY = \left\{ \prod_{j=1}^{p} a_{jj}^{j} \right\} dX,$$

$$Y = AX \Rightarrow dY = \left\{ \prod_{j=1}^{p} a_{jj}^{p+1-j} \right\} dX,$$

$$Y = aX \Rightarrow dY = a^{\frac{p(p+1)}{2}} dX, a \text{ is a real scalar constant.}$$

2.2.9. Let X, A, B be $p \times p$ lower triangular matrices where A and B are non-singular constant matrices with the diagonal elements positive and X is a matrix of distinct real variables. Then show that

$$Y = AXB \Rightarrow dY = \left\{ \prod_{j=1}^{p} a_{jj}^{j} b_{jj}^{p+1-j} \right\} dX,$$

$$Y = A'X'B' \Rightarrow dY = \left\{ \prod_{j=1}^{p} b_{jj}^{j} a_{jj}^{p+1-j} \right\} dX.$$

2.2.10. Let X be a $p \times 1$ vector of real variables and A be a $p \times p$ positive definite constant matrix. Then show that

$$\int_X e^{-X'AX} dX = \pi^{\frac{p}{2}} |A|^{-\frac{1}{2}}.$$

2.3. Nonlinear Transformations Involving Scalar Variables

One popular transformation is in terms of elementary symmetric functions. Let $x_1, \ldots, x_k$ be distinct real scalar variables and let $y_1 = x_1 + \cdots + x_k, y_2 = x_1 x_2 + \cdots + x_{k-1} x_k = $ sum of products taken two at a time, $, \ldots, y_k = x_1 x_2 \cdots x_k = $ product of all of them. It is still not known whether this transformation is one-to-one for a general k. The Jacobian can be seen to be non-zero when the x_j's are distinct. The Jacobian is given by the following.

Result 2.3.1. *In the above transformation of $x_1, \ldots, x_k$ going to the elementary symmetric functions $y_1 = \sum_{j=1}^{k} x_j, y_2 = \sum_{i \neq j} x_i x_j,$ $\ldots, y_k = \prod_{j=1}^{k} x_j$ the wedge product of differentials is*

$$dY = dy_1 \wedge \cdots \wedge dy_k = J \, dX = J \, dx_1 \wedge \cdots \wedge dx_k,$$

where

$$J = \left| \left(\frac{\partial y_i}{\partial x_j} \right) \right| = \prod_{i=1}^{k-1} \prod_{j=i+1}^{k} (x_i - x_j). \qquad (2.3.1)$$

Proof. Consider the determinant of the Jacobian matrix $\frac{\partial(y_1, \ldots, y_k)}{\partial(x_1, \ldots, x_k)}$.

$$J = \begin{vmatrix} 1 & 1 & \cdots & 1 \\ \sum_{j \neq 1, j=1}^{k} x_j & \sum_{j \neq 2, j=1}^{k} x_j & \cdots & \sum_{j \neq k, j=1}^{k} x_j \\ \vdots & \vdots & \cdots & \vdots \\ \prod_{j \neq 1, j=1}^{k} x_j & \prod_{j \neq 2, j=1}^{k} x_j & \cdots & \prod_{j \neq k, j=1}^{k} x_j \end{vmatrix}.$$

Add $-(\sum_{j \neq 1, j=1}^{k} x_j)$ times the first row to the second row etc. To make all elements in the first column, except the first one, zeros. Then use the first column to wipe out all elements in the first row except the first one. Then $x_1 - x_2$ will be a common factor in the second column, $\ldots$, $x_1 - x_k$ will be a common factor in the last column. Take out these. Now, start with the second row and repeat the process. Then $(x_2 - x_3)(x_2 - x_4) \ldots (x_2 - x_k)$ will come out, and so on.

Example 2.3.1. Verify Result 2.3.1 for $k = 3$.

Solution 2.3.1. $y_1 = x_1 + x_2 + x_3, y_2 = x_1 x_2 + x_1 x_3 + x_2 x_3, y_3 = x_1 x_2 x_3$. Then

$$\left| \left(\frac{\partial y_i}{\partial x_j} \right) \right| = \begin{vmatrix} 1 & 1 & 1 \\ x_2 + x_3 & x_1 + x_3 & x_1 + x_2 \\ x_2 x_3 & x_1 x_3 & x_1 x_2 \end{vmatrix}.$$

Add $(-1)(x_2 + x_3)$ times the first row to the second row, $(-1)(x_2 x_3)$ times the first row to the third row. The first column elements, except the first one, become zeros. Then use the first column to wipe out the second and third elements in the first row. Then we have

$$\left| \left(\frac{\partial y_i}{\partial x_j} \right) \right| = \begin{vmatrix} 1 & 0 & 0 \\ 0 & x_1 - x_2 & x_1 - x_3 \\ 0 & x_3(x_1 - x_2) & x_2(x_1 - x_3) \end{vmatrix}$$

$$= (x_1 - x_2)(x_1 - x_3) \begin{vmatrix} 1 & 0 & 0 \\ 0 & 1 & 1 \\ 0 & x_3 & x_2 \end{vmatrix}$$

$$= (x_1 - x_2)(x_1 - x_3)(x_2 - x_3).$$

The result for $k = 3$ is verified.

Another very popular transformation is the general polar coordinate transformation. Let $x_1, \ldots, x_k$ be distinct real variables, $-\infty < x_j < \infty, j = 1, \ldots, k$. Let

$$x_1 = r \sin \theta_1,$$

$$x_j = r \cos \theta_1 \cos \theta_2 \ldots \cos \theta_j \sin \theta_j, \quad j = 2, \ldots, k - 1,$$

$$x_k = r \cos \theta_1 \cos \theta_2 \ldots \cos \theta_{k-1} \tag{2.3.2}$$

for $r > 0, -\frac{\pi}{2} < \theta_j < \frac{\pi}{2}, j = 1, \ldots, k - 2, -\pi < \theta_{k-1} \leq \pi$. Then we have the following result.

Result 2.3.2. *For the above transformation in* (2.3.2)

$$dx_1 \wedge \cdots \wedge dx_k = r^{k-1} \left\{ \prod_{j=1}^{k-1} |\cos \theta_j|^{k-j-1} \right\} dr \wedge d\theta_1 \wedge \cdots \wedge d\theta_{k-1},$$

$$\tag{2.3.3}$$

ignoring the sign.

Proof. Consider the case $k = 3$. Here we have $x_1 = r \sin \theta_1, x_2 = r \cos \theta_1 \sin \theta_2, x_3 = r \cos \theta_1 \cos \theta_2$. That is,

$$x_1^2 + x_2^2 + x_3^2 = r^2,$$
$$x_2^2 + x_3^2 = r^2 \cos^2 \theta_1,$$
$$x_3^2 = r^2 \cos^2 \theta_1 \cos^2 \theta_2.$$

Now, take the differentials on both sides. Moreover

$$2x_1 dx_1 + 2x_2 dx_2 + 2x_3 dx_3 = 2r dr,$$
$$2x_2 dx_2 + 2x_3 dx_3 = 2r dr \cos^2 \theta_1 - 2r^2 \cos \theta_1 \sin \theta_1 d\theta_1,$$
$$2x_3 dx_3 = 2r dr (\cos^2 \theta_1 \cos^2 \theta_2)$$
$$- 2r^2 \cos \theta_1 \sin \theta_1 \cos^2 \theta_2 d\theta_1$$
$$- 2r^2 \cos^2 \theta_1 \cos \theta_2 \sin \theta_2 d\theta_2.$$

Take the wedge product of the terms on the left and then on the right. Note that 2 will cancel out from both sides.

$$(x_1 dx_1 + x_2 dx_2 + x_3 dx_3) \wedge (x_2 dx_2 + x_3 dx_3) \wedge (x_3 dx_3)$$
$$= [x_1 x_2 dx_1 \wedge dx_2 + x_1 x_3 dx_1 \wedge dx_3 + x_2 x_3 dx_2 \wedge dx_3$$
$$+ x_2 x_3 dx_3 \wedge dx_2] \wedge x_3 dx_3$$
$$= x_1 x_2 x_3 \, dx_1 \wedge dx_2 \wedge dx_3.$$

Now, consider the right side. On the right side we get

$$r^5 \cos^2 \theta_1 \sin \theta_1 \cos \theta_2 \sin \theta_2 \, dr \wedge d\theta_1 \wedge d\theta_2.$$

Divide both sides by $x_1 x_2 x_3$ to get

$$dx_1 \wedge dx_2 \wedge dx_3 = r^2 \cos \theta_1 \, dr \wedge d\theta_1 \wedge d\theta_2.$$

Continue the process to establish the result because the pattern remains the same.

Result 2.3.3. *Let* $r > 0, 0 < \theta_j \leq \pi, j = 1, \ldots, k-2, 0 < \theta_{k-1} \leq 2\pi$ *and let*

$$x_1 = r \sin\theta_1 \sin\theta_2 \cdots \sin\theta_{k-2} \sin\theta_{k-1}$$
$$x_2 = r \sin\theta_1 \sin\theta_2 \cdots \sin\theta_{k-2} \cos\theta_{k-1}$$
$$x_3 = r \sin\theta_1 \sin\theta_2 \cdots \cos\theta_{k-2}$$
$$\vdots$$
$$x_{k-1} = r \sin\theta_1 \cos\theta_2$$
$$x_k = r \cos\theta_1.$$

Then, ignoring the sign,

$$\mathrm{d}x_1 \wedge \cdots \wedge \mathrm{d}x_k = r^{k-1} \left\{ \prod_{j=1}^{k-1} |\sin\theta_j|^{k-j-1} \right\} \mathrm{d}r \wedge \mathrm{d}\theta_1 \wedge \cdots \wedge \mathrm{d}\theta_{k-1}.$$

$$(2.3.4)$$

Another interesting transformation is the following.

Result 2.3.4. *Let* $x_1, \ldots, x_k$ *be distinct real scalar variables with* $x_j > 0, j = 1, \ldots, k.$ *Consider the transformation,*

$$y_1 = x_1 + \cdots + x_k$$
$$y_2 = x_1^2 + \cdots + x_k^2$$
$$\vdots$$
$$y_{k-1} = x_1^{k-1} + \cdots + x_k^{k-1}$$
$$y_k = x_1 \cdots x_k.$$

Then, ignoring the sign,

$$\mathrm{d}y_1 \wedge \cdots \wedge \mathrm{d}y_k = (k-1)! \left\{ \prod_{i=1}^{k-1} \prod_{j=i+1}^{k} |x_i - x_j| \right\} \mathrm{d}x_1 \wedge \cdots \wedge \mathrm{d}x_k.$$

$$(2.3.5)$$

This can be established by using partial derivatives and the Jacobian matrix.

Exercise 2.3

2.3.1. Prove Result 2.3.4.

2.3.2. Let x_1, x_2, x_3 be independently distributed standard normal random variables. This means that their joint density is of the form

$$f(x_1, x_2, x_3) = \frac{1}{(2\pi)^{\frac{3}{2}}} e^{-\frac{1}{2}(x_1^2 + x_2^2 + x_3^2)}, \quad -\infty < x_j < \infty, \quad j = 1, 2, 3.$$

Make the polar coordinate transformation and compute the joint density of r, θ_1, θ_2. If $g(r, \theta_1, \theta_2)$ is the joint density of r, θ_1, θ_2 then

$$f(x_1, x_2, x_3) \, dx_1 \wedge dx_2 \wedge dx_3 = g(r, \theta_1, \theta_2) \, dr \wedge d\theta_1 \wedge d\theta_2.$$

Show that $g(r, \theta_1, \theta_2)$ factorizes into products of functions of r and (θ_1, θ_2) thereby showing that r and (θ_1, θ_2) are independently distributed.

2.3.3. Let x_1 and x_2 be two real positive scalar random variables having a joint density $f(x_1, x_2)$ of the form

$$f(x_1, x_2) = 2(x_2 - x_1)e^{-(x_1 + x_2)}, \quad 0 < x_1 < x_2 < \infty$$

and $f(x_1, x_2) = 0$ elsewhere. Consider a transformation into elementary symmetric functions, that is, $y_1 = x_1 + x_2, y_2 = x_1 x_2$. If the joint density of y_1 and y_2 is $g(y_1, y_2)$, then

$$f(x_1, x_2)dx_1 \wedge dx_2 = g(y_1, y_2)dy_1 \wedge dy_2.$$

Evaluate the density $g_1(y_1)$ of y_1 and show that it is of the form

$$g_1(y_1) = 2e^{-2\sqrt{y_1}}, \quad 0 < y_1 < \infty$$

and $g_1(y_1) = 0$ elsewhere. [Hint: If $g(y_1, y_2)$ is the joint density of y_1 and y_2 then the density of y_1, also called the marginal density of y_1, is available by integrating out y_2.]

2.3.4. Let $(\mathrm{d}X)$ and $(\mathrm{d}Y)$ denote the matrices of differentials in X and Y, respectively. That is,

$$X = \begin{bmatrix} x_{11} & x_{12} \\ x_{21} & x_{22} \end{bmatrix} \Rightarrow (\mathrm{d}X) = \begin{bmatrix} \mathrm{d}x_{11} & \mathrm{d}x_{12} \\ \mathrm{d}x_{21} & \mathrm{d}x_{22} \end{bmatrix},$$

$$\mathrm{d}X = \mathrm{d}x_{11} \wedge \mathrm{d}x_{12} \wedge \mathrm{d}x_{21} \wedge \mathrm{d}x_{22},$$

$$Y = \begin{bmatrix} y_{11} & y_{12} \\ y_{21} & y_{22} \end{bmatrix} \Rightarrow (\mathrm{d}Y) = \begin{bmatrix} \mathrm{d}y_{11} & \mathrm{d}y_{12} \\ \mathrm{d}y_{21} & \mathrm{d}y_{22} \end{bmatrix},$$

$$\mathrm{d}Y = \mathrm{d}y_{11} \wedge \mathrm{d}y_{12} \wedge \mathrm{d}y_{21} \wedge \mathrm{d}y_{22},$$

$$X = \begin{bmatrix} x_{11} & x_{12} \\ x_{12} & x_{22} \end{bmatrix} \Rightarrow (\mathrm{d}X) = \begin{bmatrix} \mathrm{d}x_{11} & \mathrm{d}x_{12} \\ \mathrm{d}x_{12} & \mathrm{d}x_{22} \end{bmatrix},$$

$$\mathrm{d}X = \mathrm{d}x_{11} \wedge \mathrm{d}x_{12} \wedge \mathrm{d}x_{22},$$

$$X = \begin{bmatrix} x_{11} & 0 \\ x_{21} & x_{22} \end{bmatrix} \Rightarrow (\mathrm{d}X) = \begin{bmatrix} \mathrm{d}x_{11} & 0 \\ \mathrm{d}x_{21} & \mathrm{d}x_{22} \end{bmatrix},$$

$$\mathrm{d}X = \mathrm{d}x_{11} \wedge \mathrm{d}x_{21} \wedge \mathrm{d}x_{22}.$$

Consider the linear transformation $Y = AX$ where X and Y are $p \times 1$ vectors of distinct real scalar variables and A is a $p \times p$ non-singular constant matrix. Then show that the Jacobian in the transformation $Y = AX$ is the same as the Jacobian in the transformation $(\mathrm{d}Y) = A(\mathrm{d}X)$ where X and Y are replaced by the corresponding matrices of differentials.

2.3.5. By using the fact that $XX^{-1} = I$, where X is a $p \times p$ non-singular matrix of distinct real variables x_{ij}'s, show that

$$\frac{\partial}{\partial \theta} X^{-1} = -X^{-1} \left[\frac{\partial}{\partial \theta} X \right] X^{-1},$$

where the elements of X may or may not be functions of θ. Here $\frac{\partial}{\partial \theta} X = (\frac{\partial x_{ij}}{\partial \theta}) = $ the matrix of partial derivatives of the corresponding elements.

2.3.6. By using Exercise 2.3.5 or otherwise show that if $X = (x_{ij})$ is a $p \times p$ non-singular matrix of distinct real variables then $(\mathrm{d}X^{-1}) = -X^{-1}(\mathrm{d}X)X^{-1}$.

2.3.7. Let $Y = X^{-1}$ where X and Y are $p \times p$ non-singular matrices of distinct real variables. By using Exercise 2.3.6 or otherwise show that

$$
dY = \begin{cases}
|X|^{-2p} dX & \text{for a general } X, \\
|X|^{-(p+1)} dX & \text{for } X = X', \\
|X|^{-(p-1)} dX & \text{for } X = -X', \\
|X|^{-(p+1)} dX & \text{for } X \text{ lower or upper triangular.}
\end{cases}
$$

2.3.8. Kronecker product: Let $A = (a_{ij})$ be $m \times n$ and $B = (b_{ij})$ be $p \times q$. Then the Kronecker product, denoted by $\otimes$, is defined as follows:

$$
A \otimes B = \begin{bmatrix}
a_{11}B & a_{12}B & \cdots & a_{1n}B \\
a_{21}B & a_{22}B & \cdots & a_{2n}B \\
\vdots & \vdots & \cdots & \vdots \\
a_{m1}B & a_{m2}B & \cdots & a_{mn}B
\end{bmatrix}
$$

and

$$
B \otimes A = \begin{bmatrix}
b_{11}A & b_{12}A & \cdots & b_{1q}A \\
b_{21}A & b_{22}A & \cdots & b_{2q}A \\
\vdots & \vdots & \cdots & \vdots \\
b_{p1}A & b_{p2}A & \cdots & b_{pq}A
\end{bmatrix}.
$$

Thus, $A \otimes B \neq B \otimes A$. Evaluate the Kronecker products $A \otimes B$ and $B \otimes A$ if

$$
A = \begin{bmatrix} 1 & -1 & 0 \\ 1 & 0 & 2 \end{bmatrix}, \quad B = \begin{bmatrix} 3 & -4 \\ 4 & 5 \end{bmatrix}.
$$

2.3.9. vec(X): Let $X = (x_{ij})$ be an $m \times n$ matrix. Let the jth column of X be denoted by $x_{(j)}$. Consider the $mn \times 1$ vector formed by appending $x_{(1)}, \ldots, x_{(n)}$ and forming a long string. This vector is known as vec(X) and it is

$$\text{vec}(X) = \begin{bmatrix} x_{(1)} \\ \vdots \\ x_{(n)} \end{bmatrix}.$$

If $X = \begin{bmatrix} x_{11} & x_{12} \\ x_{21} & x_{22} \end{bmatrix}$ and $A = \begin{bmatrix} 1 & -1 & 1 \\ 2 & 3 & 5 \end{bmatrix}$ then form vec(X) and vec(A).

2.3.10. Let A be $m \times n$, X be $n \times p$, and B be $p \times q$. Consider AXB and vec(AXB). Then show that, the $mq \times 1$ vector,

$$\text{vec}(AXB) = (B \otimes A)\text{vec}(X).$$

2.3.11. Let A be $p \times p$ matrix and I be $q \times q$ identity matrix. Prove that $|I \otimes A| = $ determinant of $I \otimes A = |A|^q$.

2.3.12. Let A be $p \times p$ and B be $q \times q$. Then show that $|A \otimes B| = |A|^q |B|^p$.

2.3.13. Consider the transformation $Y = AX$ where Y and X are $m \times n$ matrices of distinct real scalar variables as elements, and A is $m \times m$ non-singular constant matrix. Then by using Exercise 2.3.12 show that $dY = |A|^n dX$. [Hint: $Y = AX \Rightarrow (dY) = A(dX)$.]

2.3.14. Let $Y = XB$ where X and Y are $m \times n$ matrices of distinct real variables as elements and let B be an $n \times n$ non-singular constant matrix. Then by using Exercise 2.3.12 show that $dY = |B|^m dX$. [Hint: $Y = XB \Rightarrow (dY) = (dX)B$.]

2.3.15. Let $Y = AXB$ where X, Y, A, B be as defined in Exercises 2.3.13 and 2.3.14. Then by using Exercise 2.3.12 show that $dY = |A|^n |B|^m dX$.

2.4. Some Nonlinear Matrix Transformations

One of the most popular nonlinear matrix transformations is the one where a real positive definite matrix $X = (x_{ij}), p \times p$, is written as TT' where T is a lower triangular matrix with positive diagonal elements.

Result 2.4.1. *Let* $X = X' > O$ *be* $p \times p$ *real positive definite. Let* $T = (t_{ij}), i \geq j$ *be a real lower triangular matrix with* $t_{jj} > 0, j = 1, \ldots, p.$ *Then*

$$X = TT' \Rightarrow dX = 2^p \left\{ \prod_{j=1}^{p} t_{jj}^{p+1-j} \right\} dT. \qquad (2.4.1)$$

Proof. It can be shown that when $t_{jj} > 0, j = 1, \ldots, p$, then the transformation $X = TT'$ is one-to-one. Let $p = 3$. Further

$$X = \begin{bmatrix} x_{11} & x_{12} & x_{13} \\ x_{12} & x_{22} & x_{23} \\ x_{13} & x_{23} & x_{33} \end{bmatrix} = \begin{bmatrix} t_{11} & 0 & 0 \\ t_{21} & t_{22} & 0 \\ t_{31} & t_{32} & t_{33} \end{bmatrix} \begin{bmatrix} t_{11} & t_{21} & t_{31} \\ 0 & t_{22} & t_{32} \\ 0 & 0 & t_{33} \end{bmatrix}$$

$$= \begin{bmatrix} t_{11}^2 & t_{11}t_{21} & t_{11}t_{31} \\ t_{11}t_{21} & t_{21}^2 + t_{22}^2 & t_{21}t_{31} + t_{22}t_{32} \\ t_{11}t_{31} & t_{21}t_{31} + t_{22}t_{32} & t_{31}^2 + t_{32}^2 + t_{33}^2 \end{bmatrix}.$$

Take x_{ij}'s in the order $x_{11}, x_{12}, x_{13}, x_{22}, x_{23}, x_{33}$ and t_{ij}'s in the order $t_{11}, t_{21}, t_{31}, t_{22}, t_{32}, t_{33}$. Then the matrix of partial derivatives or Jacobian matrix is given by the following:

	t_{11}	t_{21}	t_{31}	t_{22}	t_{32}	t_{33}
x_{11}	$2t_{11}$	0	0	0	0	0
x_{12}	$*$	t_{11}	0	0	0	0
x_{13}	$*$	$*$	t_{11}	0	0	0
x_{22}	$*$	$*$	$*$	$2t_{22}$	0	0
x_{23}	$*$	$*$	$*$	$*$	t_{22}	0
x_{33}	$*$	$*$	$*$	$*$	$*$	$2t_{33}$

In the above structure, an $*$ indicates that there is a quantity that we are not interested in because of a triangular format for the Jacobian matrix. We can see that the same is the pattern for a general p. Note that one 2 comes for each j of t_{jj}. Then t_{11} appears p times, t_{22} appears $p - 1$ times, and so on, and t_{pp} appears once.

That is,

$$J = 2^p \{t_{11}^p t_{22}^{p-1} \cdots t_{pp}\} = 2^p \left\{ \prod_{j=1}^{p} t_{jj}^{p+1-j} \right\}.$$

Hence the result.

An immediate application of this result is the evaluation of a real matrix-variate gamma integral.

Example 2.4.1. Evaluate the integral $g = \int_{X>O} |X|^{\alpha - \frac{p+1}{2}} e^{-\mathrm{tr}(X)} dX$ where $X = X' > O$ is real $p \times p$ positive definite and of distinct real elements, except for symmetry. $|X|$ denotes the determinant of X, $\mathrm{tr}(X)$ is the trace of X and the integral is over all positive definite $p \times p$ matrices. Show that g is available as $g = \Gamma_p(\alpha)$, where

$$\Gamma_p(\alpha) = \pi^{\frac{p(p-1)}{4}} \Gamma(\alpha) \Gamma\left(\alpha - \frac{1}{2}\right) \cdots \Gamma\left(\alpha - \frac{p-1}{2}\right), \quad \Re(\alpha) > \frac{p-1}{2},$$

$$(2.4.2)$$

where $\Re(\cdot)$ means the real part of $(\cdot)$. $\Gamma_p(\alpha)$ is called the real matrix-variate gamma.

Solution 2.4.1. Consider the transformation $X = TT'$ where T is lower triangular with positive diagonal elements. Then $|X| = |TT'| = |T||T'| = \prod_{j=1}^{p} t_{jj}^2$. Also,

$$\mathrm{tr}(X) = \mathrm{tr}(TT') = t_{11}^2 + (t_{21}^2 + t_{22}^2) + \cdots + (t_{p1}^2 + \cdots + t_{pp}^2).$$

From Result 2.4.1 we have $dX = 2^p \{\prod_{j=1}^{p} t_{jj}^{p+1-j}\} dT$. Therefore,

$$g = \int_{TT'>O} \left\{ \prod_{j=1}^{p} (t_{jj}^2)^{\alpha - \frac{p+1}{2}} \right\} e^{-\sum_{j=1}^{p} t_{jj}^2 - \sum_{i>j} t_{ij}^2} \left\{ 2^p \prod_{j=1}^{p} t_{jj}^{p+1-j} \right\} dT$$

$$= \left\{ \prod_{j=1}^{p} 2 \int_0^{\infty} (t_{jj}^2)^{\alpha - \frac{j}{2}} e^{-t_{jj}^2} dt_{jj} \right\} \left\{ \prod_{i>j} \int_{-\infty}^{\infty} e^{-t_{ij}^2} dt_{ij} \right\}.$$

Note that

$$2 \int_0^\infty (t_{jj}^2)^{\alpha - \frac{j}{2}} e^{-t_{jj}^2} \, dt_{jj}$$

$$= \int_0^\infty u^{(\alpha - \frac{j-1}{2}) - 1} e^{-u} \, du, \quad u = t_{jj}^2$$

$$= \Gamma \left(\alpha - \frac{j-1}{2} \right), \quad \Re(\alpha) > \frac{j-1}{2}, \quad j = 1, \ldots, p.$$

But $\Re(\alpha) > \frac{j-1}{2}, j = 1, \ldots, p \Rightarrow \Re(\alpha) > \frac{p-1}{2}$. Note that

$$\int_{-\infty}^\infty e^{-t_{ij}^2} \, dt_{ij} = \sqrt{\pi} \quad \text{and} \quad \prod_{i>j} \sqrt{\pi} = \pi^{\frac{p(p-1)}{4}}.$$

Hence we have

$$\int_{X>O} |X|^{\alpha - \frac{p+1}{2}} e^{-\text{tr}(X)} \, dX$$

$$= \pi^{\frac{p(p-1)}{4}} \Gamma(\alpha) \Gamma \left(\alpha - \frac{1}{2} \right) \cdots \Gamma \left(\alpha - \frac{p-1}{2} \right)$$

$$= \Gamma_p(\alpha), \quad \text{for } \Re(\alpha) > \frac{p-1}{2}. \tag{2.4.3}$$

Therefore (2.4.3) also produces an integral representation for $\Gamma_p(\alpha)$.

Definition 2.4.1. A *real matrix-variate gamma* is given by

$$\Gamma_p(\alpha) = \pi^{\frac{p(p-1)}{4}} \Gamma(\alpha) \Gamma \left(\alpha - \frac{1}{2} \right) \cdots \Gamma \left(\alpha - \frac{p-1}{2} \right),$$

$$\times \Re(\alpha) > \frac{p-1}{2}.$$

Definition 2.4.2. A *real matrix-variate gamma density for a $p \times p$ matrix X* is given by

$$f(X) = \begin{cases} \dfrac{|B|^\alpha}{\Gamma_p(\alpha)} |X|^{\alpha - \frac{p+1}{2}} e^{-\text{tr}(BX)}, & X = X' > O, \\ \quad B = B' > O, \ \Re(\alpha) > \dfrac{p-1}{2}, \\ 0 \quad \text{elsewhere.} \end{cases} \tag{2.4.4}$$

Definition 2.4.3. *Real matrix-variate beta function:* It is denoted as $B_p(\alpha, \beta)$ and defined as

$$B_p(\alpha, \beta) = \frac{\Gamma_p(\alpha)\Gamma_p(\beta)}{\Gamma_p(\alpha + \beta)}, \quad \Re(\alpha) > \frac{p-1}{2}, \Re(\beta) > \frac{p-1}{2}. \quad (2.4.5)$$

We can show that $B_p(\alpha, \beta)$ has the following integral representations for $\Re(\alpha) > \frac{p-1}{2}, \Re(\beta) > \frac{p-1}{2}$:

$$B_p(\alpha, \beta) = \int_{O<X<I} |X|^{\alpha - \frac{p+1}{2}} |I - X|^{\beta - \frac{p+1}{2}} \mathrm{d}X,$$

$$X = X' > O, \quad I - X > O \quad (2.4.6)$$

$$= \int_{O<Y<I} |Y|^{\beta - \frac{p+1}{2}} |I - Y|^{\alpha - \frac{p+1}{2}} \mathrm{d}Y = B_p(\beta, \alpha) \quad (2.4.7)$$

$$= \int_{U>O} |U|^{\alpha - \frac{p+1}{2}} |I + U|^{-(\alpha+\beta)} \mathrm{d}U, \quad U > O \quad (2.4.8)$$

$$= \int_{V>O} |V|^{\beta - \frac{p+1}{2}} |I + V|^{-(\alpha+\beta)} \mathrm{d}V, \quad V > O. \quad (2.4.9)$$

Example 2.4.2. Establish (2.4.6).

Solution 2.4.2. We will establish one of the results and the remaining are given as exercises. Let us consider

$$\Gamma_p(\alpha)\Gamma_p(\beta) = \left[\int_{X>O} |X|^{\alpha - \frac{p+1}{2}} \mathrm{e}^{-\mathrm{tr}(X)} \mathrm{d}X \right] \left[\int_{Y>O} |Y|^{\beta - \frac{p+1}{2}} \mathrm{e}^{-\mathrm{tr}(Y)} \mathrm{d}Y \right]$$

$$= \int_{X>O} \int_{Y>O} |X|^{\alpha - \frac{p+1}{2}} |Y|^{\beta - \frac{P+1}{2}} \mathrm{e}^{-\mathrm{tr}(X+Y)} \mathrm{d}X \wedge \mathrm{d}Y.$$

Treating the right side as a double integral, let us make the transformation $U = X + Y, V = X$. The Jacobian is easily seen to be 1 and the integral reduces to the following:

$$\Gamma_p(\alpha)\Gamma_p(\beta) = \int_U \int_V |V|^{\alpha - \frac{p+1}{2}} |U - V|^{\beta - \frac{p+1}{2}} \mathrm{e}^{-\mathrm{tr}(U)} \mathrm{d}U \wedge \mathrm{d}V.$$

Note that we can write

$$|U - V| = |U||I - U^{-\frac{1}{2}}VU^{-\frac{1}{2}}| = |U||I - W|,$$

where $W = U^{-\frac{1}{2}}VU^{-\frac{1}{2}} \Rightarrow dW = |U|^{-\frac{p+1}{2}}dV$ for fixed U. That is,

$$\Gamma_p(\alpha)\Gamma_p(\beta) = \int_{U>O} |U|^{\alpha+\beta-\frac{p+1}{2}}e^{-\text{tr}(U)}dU$$

$$\times \int_{O<W<I} |W|^{\alpha-\frac{p+1}{2}}|I - W|^{\beta-\frac{p+1}{2}}dW.$$

But the U-integral is $\Gamma_p(\alpha + \beta)$. Hence, dividing by $\Gamma_p(\alpha + \beta)$ we have

$$\frac{\Gamma_p(\alpha)\Gamma_p(\beta)}{\Gamma_p(\alpha + \beta)} = \int_{O<W<I} |W|^{\alpha-\frac{p+1}{2}}|I - W|^{\beta-\frac{p+1}{2}}dW,$$

which establishes the result. Note that for the integrals to exist the conditions $\Re(\alpha) > \frac{p-1}{2}, \Re(\beta) > \frac{p-1}{2}$ are needed.

By using a real matrix-variate beta function and the integral representations in (2.4.6)–(2.4.9) we can define type-1 and type-2 real matrix-variate beta densities.

Definition 2.4.4. *Real matrix-variate type-1 beta density:* The following two forms are real matrix-variate type-1 beta densities (the non-zero parts are given below and the function is zero outside the range given there. In all the forms given below the conditions $\Re(\alpha) > \frac{p-1}{2}, \Re(\beta) > \frac{p-1}{2}$ are essential. In statistical densities, usually the parameters are real. In this case, the conditions will be $\alpha > \frac{p-1}{2}, \beta > \frac{p-1}{2}$:

$$f_1(X) = \frac{\Gamma_p(\alpha + \beta)}{\Gamma_p(\alpha)\Gamma_p(\beta)}|X|^{\alpha-\frac{p+1}{2}}|I - X|^{\beta-\frac{p+1}{2}}, \quad O < X < I.$$

$$(2.4.10)$$

$$f_2(Y) = \frac{\Gamma_p(\alpha + \beta)}{\Gamma_p(\alpha)\Gamma_p(\beta)}|Y|^{\beta-\frac{p+1}{2}}|I - Y|^{\alpha-\frac{p+1}{2}}, \quad O < Y < I.$$

$$(2.4.11)$$

Definition 2.4.5. *Real matrix-variate type-2 beta density:* The following two forms are real matrix-variate type-2 beta densities, where only the non-zero parts are given below and they are zeros outside the range. The same conditions as above are there on α and β. Further

$$f_3(U) = \frac{\Gamma_p(\alpha+\beta)}{\Gamma_p(\alpha)\Gamma_p(\beta)}|U|^{\alpha-\frac{p+1}{2}}|I+U|^{-(\alpha+\beta)}, \quad U > O, \qquad (2.4.12)$$

$$f_4(V) = \frac{\Gamma_p(\alpha+\beta)}{\Gamma_p(\alpha)\Gamma_p(\beta)}|V|^{\beta-\frac{p+1}{2}}|I+V|^{-(\alpha+\beta)}, \quad V > O. \qquad (2.4.13)$$

With the help of statistically independently distributed real matrix-variate gamma random variables we can establish connections to type-1 and type-2 beta densities. Some of these will be given here.

Result 2.4.2. *Let X_1 and X_2 be $p \times p$ real matrix-variate gamma random variables having the densities in* (2.4.4) *with parameters* $(\alpha_1, B), (\alpha_2, B)$ *respectively, where B is the same, and statistically independently distributed. Let $U = X_1 + X_2$, $V = (X_1 + X_2)^{-\frac{1}{2}} X_1(X_1 + X_2)^{-\frac{1}{2}}$, $W = X_2^{-\frac{1}{2}} X_1 X_2^{-\frac{1}{2}}$. Then U is real matrix-variate gamma distributed with the parameters $(\alpha_1+\alpha_2, B)$, V is real matrix-variate type-1 beta distributed with parameters (α_1, α_2) and W is real matrix-variate type-2 beta distributed with parameters (α_2, α_1).*

Proof. Since X_1 and X_2 are statistically independently distributed and since they have the density in (2.4.4) with the parameters (α_1, B) and (α_2, B), the joint density of X_1 and X_2, denoted by $f(X_1, X_2)$, is the following:

$$f(X_1, X_2) = \frac{|B|^{\alpha_1+\alpha_2}}{\Gamma_p(\alpha_1)\Gamma_p(\alpha_2)}|X_1|^{\alpha_1-\frac{p+1}{2}}|X_2|^{\alpha_2-\frac{p+1}{2}}e^{-\operatorname{tr}(X_1+X_2)},$$

$$X_1 > O, \ X_2 > O$$

for $\Re(\alpha_j) > \frac{p-1}{2}, j = 1, 2, B > O$. Let us make the transformation $Y_1 = X_1 + X_2, Y_2 = X_1$. Then the Jacobian is 1 and the joint density of Y_1 and Y_2, denoted by $g(Y_1, Y_2)$, is the following:

$$g(Y_1, Y_2) = \frac{|B|^{\alpha_1+\alpha_2}}{\Gamma_p(\alpha_1)\Gamma_p(\alpha_2)}|Y_2|^{\alpha_1-\frac{p+1}{2}}|Y_1 - Y_2|^{\alpha_2-\frac{p+1}{2}}e^{-\operatorname{tr}(BY_1)},$$

$$Y_2 > O, \ Y_1 - Y_2 > O$$

$$= \frac{|B|^{\alpha_1+\alpha_2}}{\Gamma_p(\alpha_1)\Gamma_p(\alpha_2)} |Y_2|^{\alpha_1-\frac{p+1}{2}} |Y_1|^{\alpha_2-\frac{p+1}{2}}$$

$$\times |I - Y_1^{-\frac{1}{2}} Y_2 Y_1^{-\frac{1}{2}}|^{\alpha_2-\frac{p+1}{2}} e^{-\mathrm{tr}(BY_1)}.$$

Put $V = Y_1^{-\frac{1}{2}} Y_2 Y_1^{-\frac{1}{2}} \Rightarrow dV = |Y_1|^{-\frac{p+1}{2}} dY_2$ for fixed Y_1. Then the joint density of V and Y_1, denoted by $h(V, Y_1)$, is the following:

$$h(V, Y_1) = \frac{|B|^{\alpha_1+\alpha_2}}{\Gamma_p(\alpha_1)\Gamma_p(\alpha_2)} |Y_1|^{\alpha_1+\alpha_2-\frac{p+1}{2}} e^{-\mathrm{tr}(BY_1)}$$

$$\times |V|^{\alpha_1-\frac{p+1}{2}} |I - V|^{\alpha_2-\frac{p+1}{2}}$$

for $Y_1 > O, O < V < I$. Since the joint density of V and Y_1 factorizes into product of functions of V and Y_1 alone, V and Y_1 are statistically independently distributed. The density of V is available by integrating out Y_1. If it is denoted by $g_2(V)$, then obviously,

$$g_2(V) = \frac{\Gamma_p(\alpha_1 + \alpha_2)}{\Gamma_p(\alpha_1)\Gamma_p(\alpha_2)} |V|^{\alpha_1-\frac{p+1}{2}} |I - V|^{\alpha_2-\frac{p+1}{2}}, \quad O < V < I$$

for $\Re(\alpha_1) > \frac{p-1}{2}, \Re(\alpha_2) > \frac{p-1}{2}$, and then obviously the marginal density of Y_1, denoted by $g_1(Y_1)$, is available as

$$g_1(Y_1) = \frac{|B|^{\alpha_1+\alpha_2}}{\Gamma_p(\alpha_1 + \alpha_2)} |Y_1|^{\alpha_1+\alpha_2-\frac{p+1}{2}} e^{-\mathrm{tr}(BY_1)}, \quad B > O, \quad Y_1 > O$$

for $\Re(\alpha_1 + \alpha_2) > \frac{p-1}{2}$. Note that

$$V = (X_1 + X_2)^{-\frac{1}{2}} X_1 (X_1 + X_2)^{-\frac{1}{2}}$$

$$= [W + I]^{-\frac{1}{2}} W [W + I]^{-\frac{1}{2}},$$

$$W = X_2^{-\frac{1}{2}} X_1 X_2^{-\frac{1}{2}} = [I + W^{-1}]^{-\frac{1}{2}} [I + W^{-1}]^{-\frac{1}{2}}$$

$$= [I + W^{-1}]^{-1} \Rightarrow V^{-1} = I + W^{-1}$$

$$\Rightarrow |V|^{-(p+1)} dV = |W|^{-(p+1)} dW \Rightarrow dV = |I + \dot{W}|^{-(p+1)} dW.$$

Then

$$|V|^{\alpha_1 - \frac{p+1}{2}}|I - V|^{\alpha_2 - \frac{p+1}{2}}\mathrm{d}V$$

$$= |I + W|^{-\alpha_1 + \frac{p+1}{2} - (p+1)}|I - (I + W^{-1})^{-1}|^{\alpha_2 - \frac{p+1}{2}}\mathrm{d}W$$

$$= |W|^{\alpha_2 - \frac{p+1}{2}}|I + W|^{-(\alpha_1 + \alpha_2)}, \quad W > O.$$

Therefore W is distributed as a real matrix-variate type-2 beta with the parameters (α_2, α_1). Hence the result. Note that $Z = W^{-1}$ is distributed as a real matrix-variate type-2 beta with the parameters (α_1, α_2).

Note that Result 2.4.2 is a very important result which establishes the connection among, gamma, type-1 and type-2 beta random variables in the real $p \times p$ matrix-variate cases. The matrix-variate gamma density given in (2.4.4) is a very important density. The main density in multivariate statistical analysis is called the Wishart density which is a particular case of (2.4.4) for

$$\alpha = \frac{n}{2}, n = p, p+1, \ldots, B = \frac{1}{2}C^{-1}, C = C' > O$$

and written as $W_p(n, C)$ or a real Wishart density for $p \times p$ matrix-variate case with degrees of freedom n and parameter matrix or covariance matrix $C > O$.

In order to show that if X is type-2 beta distributed with parameters (α, β) then $Y = X^{-1}$ is distributed as type-2 beta with parameters (β, α) we need the Jacobian of the transformation of a non-singular X going to its inverse $Y = X^{-1}$. The Jacobian is established in Exercise 2.3.7. This will be restated here as a result.

Result 2.4.3. *Let X be a $p \times p$ real non-singular matrix of distinct real scalar variables x_{ij}'s. Let $Y = X^{-1}$. Then, ignoring the sign,*

$$\mathrm{d}Y = \begin{cases} |X|^{-2p}\mathrm{d}X & \text{for a general } X, \\ |X|^{-(p+1)}\mathrm{d}X & \text{for } X = X', \\ |X|^{-(p-1)}\mathrm{d}X & \text{for } X = -X', \\ |X|^{-(p+1)}\mathrm{d}X & \text{for } X \text{ lower or upper triangular.} \end{cases}$$

Exercise 2.4

2.4.1. Let $X = X' > O$ be real $p \times p$ positive definite matrix. Consider the one to one transformation $X = TT'$ where $T = (t_{ij})$ is an upper triangular matrix with positive diagonal elements, $t_{jj} > 0$, $j = 1, \ldots, p$. Then show that

$$\mathrm{d}X = 2^p \left\{ \prod_{j=1}^{p} t_{jj}^j \right\} \mathrm{d}T$$

by using (a) the method of Section 2.4, (b) by using the fact that $(\mathrm{d}X) = (\mathrm{d}T)T' + T(\mathrm{d}T')$ and then taking the wedge product, (c) by using Kronecker product and vec notations of Exercises 2.3.8 and 2.3.9.

2.4.2. Let $X = X' > O$ be $p \times p$ real. Let $T = (t_{ij})$ and let $X = TT'$. If T is lower triangular or upper triangular show that the transformation $X = TT'$ is one-to-one when $t_{jj} > 0, j = 1, \ldots, p$.

2.4.3. Let $X = X' > O$ be a $p \times p$ real positive definite matrix of distinct real variables as elements. Let $T = (t_{ij})$ be a lower triangular matrix with positive diagonal elements such that $\sum_{j=1}^{i} t_{ij}^2 = 1, i = 1, \ldots, p$. Then show that

$$X = TT' \Rightarrow \mathrm{d}X = \left\{ \prod_{j=2}^{p} t_{jj}^{p-j} \right\} \mathrm{d}T.$$

2.4.4. For the same X and T in Exercise 2.4.3 with $\sum_{i=j}^{p} t_{ij}^2 = 1, j = 1, \ldots, p$ show that

$$X = TT' \Rightarrow \mathrm{d}X = \left\{ \prod_{j=1}^{p-1} t_{jj}^{j-1} \right\} \mathrm{d}T.$$

2.4.5. Show that for $X = X' > O, B = B' > O$ and $p \times p$ where B is a constant matrix and X is of distinct real elements, then

$$\int_{X>O} |X|^{\alpha - \frac{p+1}{2}} e^{-\mathrm{tr}(BX)} \mathrm{d}X = |B|^{-\alpha} \Gamma_p(\alpha), \quad \text{for } \Re(\alpha) > \frac{p-1}{2}$$

so that

$$|B|^{-\alpha} \equiv \frac{1}{\Gamma_p(\alpha)} \int_{X>O} |X|^{\alpha - \frac{p+1}{2}} e^{-\mathrm{tr}(BX)} dX.$$

Hint: $\mathrm{tr}(BX) = \mathrm{tr}(B^{\frac{1}{2}} X B^{\frac{1}{2}})$. Make the transformation $Y = B^{\frac{1}{2}} X B^{\frac{1}{2}}$.

2.4.6. Let X, A, B be $p \times p$ non-singular matrices where A and B are constant matrices, and X is of distinct real variables. Then show that

$$Y = AX^{-1}B \Rightarrow dY = \begin{cases} |AB|^p |X|^{-2p} dX & \text{for a general } X, \\ |AX^{-1}|^{p+1} dX & \text{for } X = X', B = A', \\ |AX^{-1}|^{p-1} dX & \text{for } X = -X', B = A'. \end{cases}$$

2.4.7. Let X and A be $p \times p$ matrices where A is a non-singular constant matrix, $A + X$ is non-singular, X of distinct real variables. Then show that

$$Y = (A + X)^{-1}(A - X) \quad \text{or} \quad (A - X)(A + X)^{-1}$$

$$\Rightarrow dY = \begin{cases} 2^{p^2} |A|^p |A + X|^{-2p} X & \text{for a general } X, \\ 2^{\frac{p(p+1)}{2}} |I + X|^{-(p+1)} dX & \text{for } A = I, X = X'. \end{cases}$$

2.4.8. Evaluate the following integrals, where X and B are $p \times p$ real positive definite:

$$I_1 = \int_{X>O} e^{-\mathrm{tr}(X)} dX; \quad I_2 = \int_{X>O} e^{-\mathrm{tr}(BX)} dX,$$

$$I_3 = \int_{X>O} |X| e^{-\mathrm{tr}(X)} dX, \quad I_4 = \int_{X>O} |X|^2 e^{-\mathrm{tr}(BX)} dX, \quad B > O.$$

2.4.9. Let $O < X < I, X = X' > O$. Let $Y = (I-X)^{-\frac{1}{2}} X (I-X)^{-\frac{1}{2}}$, $U = (I+Y)^{-\frac{1}{2}} Y (I+Y)^{-\frac{1}{2}}, V = Y^{-1}, Z = I - X$. Then show that $O < Z < I, Y = Y' > O, O < U < I, V = V' > O$.

2.4.10. By using Example 2.4.2 and Exercise 2.4.9 or otherwise establish the results in (2.4.6), (2.4.7), (2.4.8).

2.4.11. Evaluate the following integrals where X is a $p \times p$ real positive definite matrix such that $O < X < I$:

$$\text{(i):} \quad \int_{O<X<I} |X|\mathrm{d}X, \quad \text{(ii):} \quad \int_{O<X<I} \mathrm{d}X,$$

$$\text{(iii):} \quad \int_{O<X<I} |I - X|\mathrm{d}X, \quad \text{(iv):} \quad \int_{O<X<I} |X - X^2|\mathrm{d}X.$$

2.4.12. Let $X = X' > O$ be $p \times p$ real positive definite. Let $Y = X^2$ with the eigenvalues of X as $\lambda_1, \ldots, \lambda_p$. Then show that

$$\mathrm{d}Y = \begin{cases} \left\{ \displaystyle\prod_{j=1}^{p}\prod_{i=1}^{p}(\lambda_i + \lambda_j) \right\} \mathrm{d}X & \text{for a general } X, \\[3em] |X| \left\{ \displaystyle\prod_{i\neq j=1}^{p} (\lambda_i + \lambda_j) \right\} \mathrm{d}X & \text{for } X = X', \\[3em] 2^p |X| \left\{ \displaystyle\prod_{i<j=1}^{p} (\lambda_i + \lambda_j) \right\} \mathrm{d}X & \text{for } X = X'. \end{cases}$$

2.4.13. Let $Y = T^2$ where T is lower or upper triangular matrix with positive diagonal elements and $p \times p$, $t_{jj} > 0, j = 1, \ldots, p$. Then show that

$$\mathrm{d}Y = \left\{ \prod_{i=1}^{p}\prod_{j=i}^{p}(t_{ii} + t_{jj}) \right\} \mathrm{d}T.$$

2.4.14. Let X and A be symmetric $p \times p$, real and positive definite matrices where A is a constant matrix, $X = (x_{ij})$, x_{ij}'s being distinct, except for symmetry, with eigenvalues $\lambda_1 > \lambda_2 > \cdots > \lambda_p > 0$. Then show that

$$Y = XAX \Rightarrow \mathrm{d}Y = \left\{ \prod_{i\leq j}(\lambda_i + \lambda_j) \right\} \mathrm{d}X.$$

2.4.15. Let X be $p \times p$ and non-singular with inverse, determinant and cofactor of x_{ij} denoted by $X^{-1}, |X|$ and $|X_{ij}|$, respectively. Let

x_{ij}'s be distinct, except for symmetry. Then show that the differential of $|X|$ can be written as

$$d|X| = \begin{cases} \displaystyle\sum_{i=1}^{p}\sum_{j=1}^{p} |X_{ij}|dx_{ij} & \text{for a general } X, \\[2em] \displaystyle\sum_{i=1}^{p} |X_{ii}|dx_{ii} + \sum_{i>j} |X_{ij}|dx_{ij} & \text{for } X = X'. \end{cases}$$

2.4.16. Let T be $p \times p$ real, lower or upper triangular matrix with positive diagonal elements. Let $X = \frac{T}{|T|}$. Then show that

$$dY = (p-1)|T|^{-\frac{p(p+1)}{2}} dT$$

and

$$Y = X^{-1} \Rightarrow dY = (p-1)|T|^{\frac{(p+1)(p-2)}{2}} dT.$$

2.4.17. When X is real $p \times p$ matrix, show that the transformations (i): $Y = X^2$ for a general X, and (ii): $Y = X^2$ when X is skew symmetric are not one-to-one.

2.4.18. Let $X = (x_{ij})$ be a real $p \times p$ non-singular matrix with the inverse, determinant and cofactor of x_{ij} denoted by $X^{-1}, |X|$ and $|X_{ij}|$, respectively. Let $d|X|$ be the differential of $|X|$. Then show that

$$d|X| = |X|[\text{tr}[X^{-1}(dX)].$$

2.4.19. Let X be $p \times p$ real matrix of p^2 functionally independent variables. Then show that $S_1 = \frac{1}{2}(X + X')$ and $S_2 = \frac{1}{2}(X - X')$ are not one-to-one transformations but $S_1 + S_2 = X$.

2.4.20. From Exercise 2.4.19 or otherwise, show that

$$dX = 2^{\frac{p(p-1)}{2}} dS_1 \wedge dS_2,$$

where S_1, S_2 and X are as defined in Exercise 2.4.19.

2.5. Transformations Involving Orthonormal and Semiorthonormal Matrices

An orthonormal matrix V is such that $VV' = I, V'V = I$. If only one or the conditions either $VV' = I$ or $V'V = I$ hold then V is semiorthonormal. For example, let

$$V = \begin{bmatrix} \dfrac{1}{\sqrt{3}} & \dfrac{1}{\sqrt{3}} & \dfrac{1}{\sqrt{3}} \\ \dfrac{1}{\sqrt{2}} & 0 & -\dfrac{1}{\sqrt{2}} \\ \dfrac{1}{\sqrt{6}} & -\dfrac{2}{\sqrt{6}} & \dfrac{1}{\sqrt{6}} \end{bmatrix} \quad \text{and} \quad U = \begin{bmatrix} \dfrac{1}{\sqrt{3}} & \dfrac{1}{\sqrt{3}} & \dfrac{1}{\sqrt{3}} \\ \dfrac{1}{\sqrt{2}} & 0 & -\dfrac{1}{\sqrt{2}} \end{bmatrix}.$$

Then $VV' = I, V'V = I$ whereas $UU' = I, U'U \neq I$. Hence V is orthonormal and U is semiorthonormal. If V is orthonormal and $p \times p$ and if $v_{(1)}, \dots, v_{(p)}$ are the columns of V then $v'_{(i)} v_{(i)} = 1, i = 1, \dots, p$. There are p restrictions here. $v'_{(i)} v_{(j)} = 0, i \neq j, i, j = 1, \dots, p$. There are $\frac{p(p-1)}{2}$ restrictions here. Thus a total of $\frac{p(p+1)}{2}$ restrictions are there. Hence the number of free variables is $p^2 - \frac{p(p+1)}{2} = \frac{p(p-1)}{2}$. If U is $q \times p, q \geq p$ and if the p columns of U are orthonormal then $U'U = I_p$ but $UU' \neq I$. Then the total number of restrictions on U is $\frac{p(p+1)}{2}$ and the total number of free variables in U is $pq - \frac{p(p+1)}{2}$. If V is orthonormal and if any row or column is multiplied by (-1) still the new matrix is orthonormal or the orthonormality is not lost. These observations will be made use of in the coming discussions.

When $X = X'$ is $p \times p$ then we know that there exists an orthonormal matrix V such that $V'XV = D = \text{diag}(\lambda_1, \dots, \lambda_p)$ or $X = VDV'$ where D is a diagonal matrix with the diagonal elements being the eigenvalues of X. The eigenvalues of a symmetric matrix are real. The orthonormal matrix V here is not unique as indicated above. We will examine the Jacobian of this transformation $X = VDV'$ next.

Result 2.5.1. *Let* $X = (x_{ij})$ *be* $p \times p$ *real symmetric matrix of distinct real variables* x_{ij}'s*. Let* $\lambda_1, \dots, \lambda_p$ *be the eigenvalues of* X *and let them be distinct. Let* V *be a unique orthonormal matrix,*

$VV' = I, V'V = I$ *such that* $V'XV = D = \text{diag}(\lambda_1, \ldots, \lambda_p)$. *Let* $\lambda_1 > \lambda_2 > \cdots > \lambda_p$, $\lambda_j \neq 0, j = 1, \ldots, p$. *Then* $X = VDV'$, *and ignoring the sign,*

$$\mathrm{d}X = \left\{ \prod_{i=1}^{p-1} \prod_{j=i+1}^{p} |\lambda_i - \lambda_j| \right\} \mathrm{d}D \wedge \mathrm{d}G,$$

where the matrix of differentials in G, *namely* $(\mathrm{d}G) = V'(\mathrm{d}V)$.

Proof. Consider $X = VDV'$ and take the matrix of differentials. That is,

$$(\mathrm{d}X) = (\mathrm{d}V)DV' + V(\mathrm{d}D)V' + VD(\mathrm{d}V').$$

Premultiply by V' and postmultiply by V to get

$$\text{(i)} \quad V'(\mathrm{d}X)V = V'(\mathrm{d}V)D + (\mathrm{d}D) + D(\mathrm{d}V')V.$$

Let $(\mathrm{d}W) = V'(\mathrm{d}X)V, (\mathrm{d}Y) = V'(\mathrm{d}V)$. Then

$$\text{(ii)} \quad (\mathrm{d}W) = (\mathrm{d}Y)D + (\mathrm{d}D) + D(\mathrm{d}Y).$$

Since orthonormal matrices have determinant ± 1 we have $|V| = \pm 1$ and the wedge product of differentials in W is $\mathrm{d}W = |V|^{p+1}\mathrm{d}X = \mathrm{d}X$, ignoring the sign. But from $V'V = I$ we have $V'(\mathrm{d}V) + (\mathrm{d}V')V = O \Rightarrow V'(\mathrm{d}V) = -(\mathrm{d}V')V = -[V'(\mathrm{d}V)]'$ or $(\mathrm{d}Y) = V'(\mathrm{d}V)$ is a skew symmetric matrix having $y_{jj} = 0, j = 1, \ldots, p$ and $y_{ij} = -y_{ji}, i > j$. Let us look at the diagonal elements in $(\mathrm{d}W)$. Since $(\mathrm{d}Y)$ is skew symmetric we have from (ii)

$$\text{(iii)} \quad \mathrm{d}w_{ii} = \mathrm{d}\lambda_i \Rightarrow \bigwedge_{i=1}^{p} \mathrm{d}w_{ii} = \bigwedge_{i=1}^{p} \mathrm{d}\lambda_i = \mathrm{d}D.$$

Now, consider $\mathrm{d}w_{ij}$ for $i > j$. From (ii), $\mathrm{d}w_{ij} = (\lambda_j - \lambda_i)\mathrm{d}y_{ij}, i > j$. Hence

$$\text{(iv)} \quad \bigwedge_{i>j=1}^{p} \mathrm{d}w_{ij} = \left\{ \prod_{i>j}(\lambda_j - \lambda_i) \right\} \mathrm{d}Y, \qquad (\mathrm{d}Y) = V'(\mathrm{d}V).$$

Hence from (iii) and (iv) and the fact that $dX = dW$, ignoring the sign, we have, ignoring the sign

(v) $\quad dX = dW = \left\{ \prod_{i \geq j=1}^{p} |\lambda_i - \lambda_j| \right\} dD \wedge dY, \quad (dY) = V'(dV).$

This establishes the result.

This is the most important transformation involving a symmetric matrix X and a full orthonormal matrix V. But Eq. (v) has a problem of integration over Y. What is the value of the integral over Y of $V'(dV)$ or $\int_Y V'(dV)$? Due to the presence of the factor $\prod_{i \geq j=1}^{p} |\lambda_i - \lambda_j|$ it will be difficult to evaluate $\int_Y dY, (dY) = V'(dV)$, by going through the above result. To this end we will consider another transformation involving a triangular matrix and an orthonormal matrix.

Let X be a $p \times p$ matrix of p^2 real distinct variables. Let $T = (t_{ij})$ be a lower triangular matrix with diagonal elements positive. Let V be an orthonormal matrix $VV' = I, V'V = I$. One can restrict the diagonal elements of V to be positive to select a unique V. In V there are $\frac{p(p-1)}{2}$ distinct variables and in T we have $\frac{p(p+1)}{2}$ distinct variables, giving a total of $\frac{p(p+1)}{2} + \frac{p(p-1)}{2} = p^2$ variables. Then we can consider the transformation $X = TV$.

Result 2.5.2. *Let $X = (x_{ij})$ be real $p \times p$ matrix of p^2 distinct real variables. Let $T = (t_{ij}), i \geq j$, be a lower triangular matrix with positive diagonal elements and distinct real t_{ij}'s, $i > j$. Let V be a unique orthonormal matrix, $VV' = I_p, V'V = I_p$. Then*

$$X = TV' \Rightarrow dX = \left\{ \prod_{j=1}^{p} t_{jj}^{p-j} \right\} dT \wedge dY, \quad (dY) = V'(dV).$$

Proof. Take the matrix of differentials in $X = TV$.

$$(dX) = (dT)V' + T(dV').$$

Postmultiply by V to get

(i) $\quad (dX)V = (dT) + T(dV')V.$

Let $(\mathrm{d}W) = (\mathrm{d}X)V$ and let $(\mathrm{d}Y) = V'(\mathrm{d}V)$. Then

$$\text{(ii)} \quad (\mathrm{d}W) = (\mathrm{d}T) + T(\mathrm{d}Y').$$

Since X is a general $p \times p$ matrix, $(\mathrm{d}W) = (\mathrm{d}X)V \Rightarrow \mathrm{d}W = |V|^p \mathrm{d}X = \mathrm{d}X$, ignoring the sign, since $|V| = \pm 1$. Let $(\mathrm{d}U) = T(\mathrm{d}Y')$. Observe that

$$(\mathrm{d}U) = T(\mathrm{d}Y') = \begin{bmatrix} t_{11} & 0 & \cdots & 0 \\ t_{21} & t_{22} & \cdots & 0 \\ \vdots & \vdots & \cdots & \vdots \\ t_{p1} & t_{p2} & \cdots & t_{pp} \end{bmatrix} \begin{bmatrix} 0 & \mathrm{d}y_{21} & \cdots & \mathrm{d}y_{p1} \\ -\mathrm{d}y_{21} & 0 & \cdots & \mathrm{d}y_{p2} \\ \vdots & \vdots & \cdots & \vdots \\ -\mathrm{d}y_{p1} & -\mathrm{d}y_{p2} & \cdots & 0 \end{bmatrix}.$$

When taking wedge product in $(\mathrm{d}U)$ we have t_{11} coming $p-1$ times, t_{22} coming $p-2$ times and so on. Then, ignoring the sign,

$$\mathrm{d}U = \left\{ \prod_{j=1}^{p-1} t_{jj}^{p-j} \right\} \mathrm{d}Y.$$

Therefore, taking wedge product in $(\mathrm{d}W)$ we have

$$\mathrm{d}W = \mathrm{d}T \wedge \left\{ \prod_{j=1}^{p-1} t_{jj}^{p-j} \right\} \mathrm{d}Y$$

$$\Rightarrow \mathrm{d}X = \left\{ \prod_{j=1}^{p-1} t_{jj}^{p-j} \right\} \mathrm{d}T \wedge \mathrm{d}Y, (\mathrm{d}Y) = V'(\mathrm{d}V).$$

This completes the proof.

Let us evaluate $\int_Y \mathrm{d}Y, (\mathrm{d}Y) = V'(\mathrm{d}V)$. To this end, consider a $p \times p$ general matrix X and consider the following integral:

$$\int_X \mathrm{e}^{-\mathrm{tr}(XX')} \mathrm{d}X = \int_X \mathrm{e}^{-\sum_{i,j} x_{ij}^2} \mathrm{d}X = \prod_{i,j} \int_{-\infty}^{\infty} \mathrm{e}^{-x_{ij}^2} \mathrm{d}x_{ij}$$

$$= \prod_{i,j} \sqrt{\pi} = \pi^{\frac{p^2}{2}}.$$

Now, let us evaluate the same integral by using the transformation $X = TV$ where T is lower triangular with positive diagonal elements and V is an orthonormal matrix.

$$\int_X e^{-\mathrm{tr}(XX')} \mathrm{d}X = \int_{TV} e^{-\mathrm{tr}[(TV)(TV)']} \mathrm{d}X$$

$$= \int_T e^{-\mathrm{tr}(TT')} \left\{ \prod_{j=1}^p t_{jj}^{p-j} \right\} \mathrm{d}T \wedge \mathrm{d}Y, \ (\mathrm{d}Y) = V'(\mathrm{d}V).$$

Note that,

$$\mathrm{tr}(TT') = t_{11}^2 + (t_{21}^2 + t_{22}^2) + \cdots + (t_{p1}^2 + \cdots + t_{pp}^2)$$

and

$$\int_T e^{-\mathrm{tr}(TT')} \mathrm{d}T = \left\{ \prod_{j=1}^p \int_0^\infty t_{jj}^{p-j} e^{-t_{jj}^2} \mathrm{d}t_{jj} \right\} \left\{ \prod_{i>j} \int_{\infty}^\infty e^{-t_{ij}^2} \mathrm{d}t_{ij} \right\}$$

$$= \left\{ \prod_{j=1}^p 2^{-1} \int_0^\infty u^{\frac{p}{2} - \frac{i-1}{2} - 1} e^{-u} \mathrm{d}u, u = t_{jj}^2 \right\} \left\{ \prod_{i>j} \sqrt{\pi} \right\}$$

$$= 2^{-p} \pi^{\frac{p(p-1)}{4}} \prod_{j=1}^p \Gamma\left(\frac{p}{2} - \frac{j-1}{2}\right) = 2^{-p} \Gamma_p\left(\frac{p}{2}\right).$$

since

$$\int_{-\infty}^\infty e^{-t_{ij}^2} \mathrm{d}t_{ij} = \sqrt{\pi} \quad \text{and} \quad \int_0^\infty t_{jj}^{p-j} e^{-t_{jj}^2} \mathrm{d}t_{jj} = \frac{1}{2} \Gamma\left(\frac{p}{2} - \frac{j-1}{2}\right)$$

for $j = 1, \ldots, p$. Therefore, for $(\mathrm{d}Y) = V'(\mathrm{d}V)$

$$\int_Y \mathrm{d}Y = \frac{2^p \pi^{\frac{p^2}{2}}}{\Gamma_p(\frac{p}{2})}.$$

This establishes the result.

Result 2.5.3. *Let V be an orthonormal matrix, $VV' = I, V'V = I$. Let the matrix of differential* $(\mathrm{d}Y) = V(\mathrm{d}V')$ *where* $(\mathrm{d}V')$ *is the* *matrix of differentials in* V'. *Then for* $(\mathrm{d}Y) = V(\mathrm{d}V')$

$$\int_Y \mathrm{d}Y = \frac{2^p \pi^{\frac{p^2}{2}}}{\Gamma_p(\frac{p}{2})}.$$

We can extend Results 2.5.2 and 2.5.3 to rectangular matrices and semiorthonormal matrices. The proofs are parallel and hence we will state the results without proofs.

Result 2.5.4. *Let X be a $p \times n$, $n \geq p$ matrix of rank p and let T be a $p \times p$ lower triangular matrix with positive diagonal elements. Let U be an $n \times p$ semiorthonormal matrix $U'U = I_p$, all are of functionally independent real scalar variables. Then*

$$X = TU' \Rightarrow \mathrm{d}X = \left\{ \prod_{j=1}^{p} t_{jj}^{n-j} \right\} \mathrm{d}T \wedge \mathrm{d}Y, \quad (\mathrm{d}Y) = U(\mathrm{d}U').$$

Result 2.5.5. *Let X, U, Y be as defined in Result 2.5.4. Then*

$$\int_Y \mathrm{d}Y = \frac{2^p \pi^{\frac{np}{2}}}{\Gamma_p(\frac{n}{2})}, \quad (\mathrm{d}Y) = U(\mathrm{d}U').$$

Exercise 2.5

2.5.1. Prove Result 2.5.4. Hint: Take additional submatrix so that U augmented with an additional submatrix creates a full orthonormal matrix. Then use Result 2.5.2.

2.5.2. Prove Result 2.5.5. Hint: Proceed as in the proof of Result 2.5.3.

2.5.3. Let $X = (x_{ij})$ be $p \times n, n \geq p$ matrix of distinct real variables x_{ij}'s. Let T be a $p \times p$ real lower triangular matrix with positive diagonal elements. Let V_1 be an $n \times p$ semiorthonormal matrix, $V_1'V_1 = I_p$. Consider $X = TV_1'$ and let $S = XX' = TT'$. Then show that

$$\mathrm{d}X = 2^{-p}|S|^{\frac{n}{2} - \frac{p+1}{2}} \mathrm{d}S \wedge \mathrm{d}Y, \quad (\mathrm{d}Y) = V_1(\mathrm{d}V_1')$$

and that

$$\int_Y dY = \frac{2^p \pi^{\frac{np}{2}}}{\Gamma_p(\frac{n}{2})}.$$

2.5.4. Let $X = (x_{ij})$ and $T = (t_{ij})$ be $p \times p$ lower triangular matrices of functionally independent real variables where the diagonal elements of T are unities. Let D be a diagonal matrix with real, non-zero and distinct diagonal elements $\lambda_1, \ldots, \lambda_p$. Then show that

$$X = DT \Rightarrow dX = \left\{ \prod_{j=1}^p |\lambda_j|^{j-1} \right\} dD \wedge dT;$$

$$X = TD \Rightarrow dX = \left\{ \prod_{j=1}^p |\lambda_j|^{p-j}| \right\} dD \wedge dT.$$

2.5.5. In Exercise 2.5.4 let $\lambda_j > 0, j = 1, \ldots, p$. Then show that

$$Y = D^{\frac{1}{2}}T \Rightarrow dY = \left\{ 2^{-p} \prod_{j=1}^p (\lambda_j^{\frac{1}{2}})^{j-2} \right\} dD \wedge dT$$

$$Y = TD^{\frac{1}{2}} \Rightarrow dY = \left\{ 2^{-p} \prod_{j=1}^p (\lambda_j^{\frac{1}{2}})^{p-1-j} \right\} dD \wedge dT,$$

where $D^{\frac{1}{2}} = \text{diag}(\lambda_1^{\frac{1}{2}}, \ldots, \lambda_p^{\frac{1}{2}})$.

2.5.6. Let X and T be as defined in Exercise 2.5.4 with $\lambda_j > 0, j = 1, \ldots, p$. Then show that

$$X = TDT' \Rightarrow dX = \left\{ \prod_{j=1}^p \lambda_j^{p-j} \right\} dD \wedge dT;$$

$$X = D^{\frac{1}{2}}TT'D^{\frac{1}{2}} \Rightarrow dX = \left\{ \prod_{j=1}^p \lambda_j^{\frac{p-1}{2}} \right\} dD \wedge dT;$$

$$X = T'DT \Rightarrow \mathrm{d}X = \left\{ \prod_{j=1}^{p} \lambda_j^{j-1} \right\} \mathrm{d}D \wedge \mathrm{d}T;$$

$$X = D^{\frac{1}{2}}T'TD^{\frac{1}{2}} \Rightarrow \mathrm{d}X = \left\{ \prod_{j=1}^{p} \lambda_j^{\frac{p-1}{2}} \right\} \mathrm{d}D \wedge \mathrm{d}T.$$

2.5.7. Let $X = (x_{ij})$ and $Y = (y_{ij})$ be real $p \times p$ matrices of distinct variables x_{ij}'s and y_{ij}'s with $y_{jj} > 0, j = 1, \ldots, p$ and $x_{jj} = 1$, $j = 1, \ldots, p$ and let D be a diagonal matrix with distinct diagonal elements $\lambda_j > 0, j = 1, \ldots, p$. Then show that

$$Y = DXD \Rightarrow \mathrm{d}Y = 2^p \left\{ \prod_{j=1}^{p} \lambda_j^{2p-1} \right\} \mathrm{d}X \wedge \mathrm{d}D;$$

$$Y = DX \Rightarrow \mathrm{d}Y = \left\{ \prod_{j=1}^{p} \lambda_j^{p-1} \right\} \mathrm{d}X \wedge \mathrm{d}D;$$

$$Y = XD \Rightarrow \mathrm{d}Y = \left\{ \prod_{j=1}^{p} \lambda_j^{p-1} \right\} \mathrm{d}X \wedge \mathrm{d}D;$$

$$Y = DXD \Rightarrow \mathrm{d}Y = \left\{ 2^p \prod_{j=1}^{p} \lambda_j^{p} \right\} \mathrm{d}X \wedge \mathrm{d}D.$$

2.5.8. Let $X = X'$ be $p \times p$ real positive definite matrix having the density

$$f(X) = \frac{|X|^{\alpha - \frac{p+1}{2}} \mathrm{e}^{-\mathrm{tr}(X)}}{\Gamma_p(\alpha)}, \quad \Re(\alpha) > \frac{p-1}{2}, X > O.$$

Let $X = D^{\frac{1}{2}}YD^{\frac{1}{2}}$ and $D = \mathrm{diag}(\lambda_1, \ldots, \lambda_p), \lambda_j > 0, j = 1, \ldots, p$. Let $Y = (y_{ij})$ with $y_{jj} = 1, j = 1, \ldots, p$. Evaluate the density of Y.

2.5.9. For a 3×3 real positive definite matrix X, such that $O < X < I$ show that

$$\int_O^I \mathrm{d}X = \frac{\pi^2}{90}.$$

2.5.10. Let $D = \operatorname{diag}(\lambda_1, \ldots, \lambda_p), 0 < \lambda_1 < \lambda_2 < \cdots < \lambda_p < \infty$. Let Ω be the region $\Omega = \{(\lambda_1, \ldots, \lambda_k) | 0 < \lambda_1 < \cdots < \lambda_p < \infty\}$. Establish the following results:

$$\int_\Omega \left[\prod_{j>k} (\lambda_j - \lambda_k) \right] e^{-\operatorname{tr}(D)} dD = \frac{\Gamma_p(\frac{p+1}{2})\Gamma_p(\frac{p}{2})}{\pi^{\frac{p^2}{2}}}.$$

$$\int_\Omega \left\{ \prod_{j=1}^p \lambda_j^{\alpha - \frac{p+1}{2}} \right\} \left\{ \prod_{i>j} (\lambda_i - \lambda_j) \right\} e^{-\operatorname{tr}(D)} dD = \frac{\Gamma_p(\alpha)\Gamma_p(\frac{p}{2})}{\pi^{\frac{p^2}{2}}}.$$

$$\int_{0<\lambda_1<\cdots<\lambda_p<1} \left\{ \prod_{i>j} (\lambda_i - \lambda_j) \right\} \left\{ \prod_{j=1}^p \lambda_j^{\alpha - \frac{p+1}{2}} \right\} \left\{ \prod_{j=1}^p (1 - \lambda_j)^{\beta - \frac{p+1}{2}} \right\} dD$$

$$= \frac{\Gamma_p(\alpha)\Gamma_p(\beta)\Gamma_p(\frac{p}{2})}{\Gamma_p(\alpha + \beta)\pi^{\frac{p^2}{2}}}.$$

2.6. Some Matrix Transformations in the Complex Domain

All the results on Jacobians in the real matrix case can be extended to the case of complex matrices also. Results and all the steps are parallel. Hence we will list a few basic results here without proofs. The proofs and other related matrix transformations may be seen from [3]. Consider a $p \times p$ matrix $\tilde{Z}$ in the complex domain. Variable matrices in the complex domain will be written with a tilde. Constant matrices, real or complex will be written without a tilde. Then $\tilde{Z}$ can be written as $\tilde{Z} = X + iY$ where $i = \sqrt{-1}$, X, Y are $p \times p$ real matrices. Then the conjugate of $\tilde{Z}$ is denoted by $\bar{\tilde{Z}} = X - iY$. Transposes will be written with a prime. Conjugate transposes will be denoted by an $*$. If $\tilde{Z}$ is hermitian then $\tilde{Z} = \tilde{Z}^*$, that is, $\tilde{Z}$ is equal to its conjugate transpose, then X will be symmetric and Y will be skew symmetric. $\tilde{Z}^* = X' - iY'$. When $\tilde{Z}$ is hermitian then $X + iY = X' - iY' \Rightarrow X = X', Y = -Y'$. The wedge product of differentials in $\tilde{Z}$ will be defined as

$$d\tilde{Z} = dX \wedge dY.$$

If $\tilde{z} = x + iy$ is a scalar complex variable, then $d\tilde{z} = dx \wedge dy$. Let $\det(\tilde{Z})$ denote the determinant of $\tilde{Z}$. This, when computed, will be of the form $\det(\tilde{Z}) = a + ib$ where a and b are real scalars. Then the determinant of the conjugate will be of the form $\det(\bar{\tilde{Z}}) = a - ib$. Then the absolute value of the determinant of $\tilde{Z}$, denoted by $|\det(\tilde{Z})|$ will be of the form

$$|\det(\tilde{Z})| = [(a + ib)(a - ib)]^{\frac{1}{2}} = [a^2 + b^2]^{\frac{1}{2}} = |\det(\tilde{Z}\tilde{Z}^*)|^{\frac{1}{2}}. \quad (2.6.1)$$

Jacobians of various matrix transformations can be worked out parallel to those in the real case by using the definitions and ideas given above. Due to space limitations we will only list some of the results here for the sake of reference. For proofs, see [3], the book mentioned earlier. The following observations will be helpful when deriving the results listed below. If $\tilde{z} = x + iy$ is a scalar complex variable then $d\tilde{z} = dx \wedge dy$. Then $\tilde{z}^* = $ conjugate transpose $=$ conjugate $= x - iy$. Then

$$d\tilde{z}^* = dx \wedge (-dy) = -dx \wedge dy = -d\tilde{z}.$$

If $\tilde{Z}$ is an $m \times n$ matrix of complex variables and if $\tilde{Z} = X + iY$ then the transpose $\tilde{Z}' = X' + iY'$. For real X and Y, $dX' = dX$ and $dY' = dY$. Then

$$d\tilde{Z}' = dX' \wedge dY' = dX \wedge dY = d\tilde{Z}.$$

But for an $m \times n$ matrix

$$\tilde{Z}^* = X' - iY' \Rightarrow d\tilde{Z}^* = dX' \wedge d(-Y') = dX \wedge d(-Y)$$

$$= (-1)^{mn} dX \wedge dY = (-1)^{mn} d\tilde{Z} \Rightarrow d\tilde{Z}^* = (-1)^{mn} d\tilde{Z}.$$

Example 2.6.1. Let $\tilde{z} = x + iy$ where x and y are real scalar variables. Let $a = 1 + 2i$ be a scalar constant. Let $\tilde{u} = a\tilde{z}$. Evaluate the connection between $d\tilde{u}$ and $d\tilde{z}$.

Solution 2.6.1. $dz = dx \wedge dy_r$

$$\tilde{u} = a\tilde{z} = (1 + 2i)(x + iy) = (x - 2y) + i(2x + y)$$
$$= u_1 + iu_2, u_1 = x - 2y, u_2 = 2x + y.$$

Then $d\tilde{u} = du_1 \wedge du_2$. Moreover

$$\frac{\partial(u_1, u_2)}{\partial(x, y)} = \begin{bmatrix} 2 & -2 \\ 2 & 1 \end{bmatrix} \Rightarrow du_1 \wedge du_2 = \begin{vmatrix} 1 & -2 \\ 2 & 1 \end{vmatrix} = 5 = |a|^2.$$

Hence

$$\tilde{u} = a\tilde{z} \Rightarrow du = |a|^2 d\tilde{z}.$$

Result 2.6.1. *Let $\tilde{X}$ and $\tilde{Y}$ be $p \times 1$ vectors of distinct complex variables each. Let A be a $p \times p$ non-singular constant matrix and α be a scalar constant. Let C be a constant $p \times 1$ vector. Then*

$$\tilde{Y} = A\tilde{X} + C \Rightarrow d\tilde{Y} = |\det(A)|^2 d\tilde{X} = |\det(AA^*)| d\tilde{X};$$
$$\tilde{Y} = \alpha\tilde{X} \Rightarrow d\tilde{Y} = |\alpha|^{2p} d\tilde{X};$$
$$\tilde{Y}' = \tilde{X}'A' \Rightarrow d\tilde{Y}' = d\tilde{Y} = |\det(AA^*)| d\tilde{X}; \qquad (2.6.2)$$
$$\tilde{Y}^* = \tilde{X}^*A^* \Rightarrow d\tilde{Y}^* = (-1)^p |\det(AA^*)| d\tilde{X}.$$

Result 2.6.2. *Let $\tilde{X}$ and $\tilde{Y}$ be $m \times n$ matrices of distinct complex variables each. Let A be $m \times m$ and B be $n \times n$ constant non-singular matrices. Let C be an $m \times n$ constant matrix and α an constant scalar. Then, ignoring the sign*

$$\tilde{Y} = A\tilde{X} + C \Rightarrow d\tilde{Y} = |\det(AA^*)|^n d\tilde{X};$$
$$\tilde{Y} = \tilde{X}B + C \Rightarrow d\tilde{Y} = |\det(BB^*)|^m d\tilde{X};$$
$$\tilde{Y} = A\tilde{X}B + C \Rightarrow d\tilde{X} = |\det(AA^*)|^n |\det(BB^*)|^m d\tilde{X}; \qquad (2.6.3)$$
$$\tilde{Y} = \alpha\tilde{X} + C \Rightarrow d\tilde{Y} = |\alpha|^{2mn} d\tilde{X}.$$

Result 2.6.3. *Let $\tilde{X}, \tilde{Y}, A, B, C$ be $p \times p$ lower triangular matrices where A, B, C are constant matrices and $\tilde{X}, \tilde{Y}$ are of functionally* *independent complex variables each. Let* α *be a scalar constant. Then*

$$\tilde{Y} = \tilde{X} + \tilde{X}' + C \Rightarrow d\tilde{Y} = \begin{cases} 2^{2p} d\tilde{X} & \text{for a general } \tilde{X}, \\ 2^p d\tilde{X} & \text{if } x_{jj}\text{'s are real;} \end{cases}$$

$$\tilde{Y} = A\tilde{X} + C \Rightarrow dY = \begin{cases} \left\{ \prod_{j=1}^{p} |a_{jj}|^{2j} \right\} d\tilde{X} & \text{for a general } \tilde{X}, \\ \left\{ \prod_{j=1}^{p} |a_{jj}|^{2j-1} \right\} d\tilde{X} & \text{if } a_{jj}\text{'s and } x_{jj}\text{'s} \\ & \text{are real;} \end{cases}$$

$$\tilde{Y} = \tilde{X}B + C \Rightarrow d\tilde{Y} = \begin{cases} \left\{ \prod_{j=1}^{p} |b_{jj}|^{2(p-j+1)} \right\} d\tilde{X} & \text{for a general } \tilde{X}, \\ \left\{ \prod_{j=1}^{p} |b_{jj}|^{2(p-j)+1} \right\} d\tilde{X} & \text{if } b_{jj}\text{'s and } x_{jj}\text{'s} \\ & \text{are real;} \end{cases}$$

$$\tilde{Y} = \alpha\tilde{X} + C \Rightarrow dY = |\alpha|^{p(p+1)} d\tilde{X}.$$

$$(2.6.4)$$

Result 2.6.4. *Let* $\tilde{X}, A$ *be* $p \times p$ *matrices where* A *is a non-singular constant matrix and* $\tilde{X}$ *is a hermitian matrix of distinct complex variables. Then, ignoring the sign,*

$$\tilde{Y} = A\tilde{X}A^* \Rightarrow d\tilde{Y} = |\det(AA^*)|^p d\tilde{X}. \qquad (2.6.5)$$

Result 2.6.5. *Let* $\tilde{X}$ *be* $p \times p$ *hermitian positive definite matrix of functionally independent complex variables under the restriction of positive definiteness. Let* $\tilde{T}, \tilde{U}$ *be* $p \times p$ *lower triangular and upper triangular matrices of distinct complex variables respectively with real and positive diagonal elements. Then*

$$\tilde{X} = \tilde{T}\tilde{T}^* \Rightarrow d\tilde{X} = 2^p \left\{ \prod_{j=1}^{p} t_{jj}^{2(p-j)+1} \right\} d\tilde{X};$$

$$\tilde{X} = \tilde{U}\tilde{U}^* \Rightarrow d\tilde{X} = 2^p \left\{ \prod_{j=1}^{p} t_{jj}^{2(j-1)+1} \right\} d\tilde{X}. \qquad (2.6.6)$$

One immediate application of this decomposition of a hermitian positive definite matrix into triangular format is the extension of matrix-variate gamma function to the complex domain and to obtain an integral representation for the same.

Definition 2.6.1. *Complex matrix-variate gamma function:* It is denoted with a tilde as $\tilde{\Gamma}_p(\alpha)$ and is defined as follows:

$$\tilde{\Gamma}_p(\alpha) = \pi^{\frac{p(p-1)}{2}} \Gamma(\alpha)\Gamma(\alpha-1)\cdots\Gamma(\alpha-p+1),$$

$$\Re(\alpha) > p - 1. \qquad (2.6.7)$$

It has the following integral representation, where $\tilde{X}$ is a $p \times p$ hermitian positive definite.

$$\tilde{\Gamma}_p(\alpha) = \int_{\tilde{X}>O} |\det(\tilde{X})|^{\alpha-p} e^{-\mathrm{tr}(\tilde{X})} d\tilde{X}.$$

Proof *(for the integral representation).* Let $\tilde{X} = \tilde{T}\tilde{T}^*$ where $\tilde{T}$ is a $p \times p$ lower triangular matrix with real and positive diagonal elements. Then this transformation can be shown to be one-to-one. Then from Result 2.6.6 we have

$$d\tilde{X} = 2^p \left\{ \prod_{j=1}^{p} t_{jj}^{2(p-j)+1} \right\} d\tilde{T}.$$

Note that

$$|\det(\tilde{X})| = |\det(\tilde{T}\tilde{T}^*)| = |\det(\tilde{T})|^2 = \prod_{j=1}^{p} t_{jj}^2$$

and

$$\mathrm{tr}(\tilde{X}) = \mathrm{tr}(\tilde{T}\tilde{T}^*) = t_{11}^2 + \cdots + t_{pp}^2 + \sum_{i>j} |\tilde{t}_{ij}|^2.$$

Then

$$|\det(\tilde{X})|^{\alpha-p}e^{-\text{tr}(\tilde{X})}d\tilde{X} = \left\{\prod_{j=1}^{p}(t_{jj}^{2})^{\alpha-j+\frac{1}{2}}\right\} e^{-\sum_{j=1}^{p}t_{jj}^{2}-\sum_{i>j}|\tilde{t}_{ij}|^{2}}d\tilde{T}.$$

The integral over $\tilde{X}$ factorizes into integrals over $t_{jj}, j = 1,\ldots,p$ and $|\tilde{t}_{ij}|^{2}$ for $i > j$. That is,

$$\int_{\tilde{X}>O} |\det(\tilde{X})|^{\alpha-p}e^{-\text{tr}(\tilde{X})}d\tilde{X}$$

$$= \left\{\prod_{j=1}^{p}\int_{0}^{\infty} 2(t_{jj}^{2})^{\alpha-j+\frac{1}{2}}e^{-t_{jj}^{2}}dt_{jj}\right\}\left\{\prod_{i>j}\int_{\tilde{t}_{ij}} e^{-|\tilde{t}_{ij}|^{2}}d\tilde{t}_{ij}\right\}.$$

But

$$\int_{0}^{\infty} 2(t_{jj}^{2})^{\alpha-j+\frac{1}{2}}e^{-t_{jj}^{2}}dt_{jj} = \Gamma(\alpha-j+1),$$

$$\Re(\alpha) > j-1, j = 1,\ldots,p$$

or $\Re(\alpha) > p-1$. Note that for $\tilde{t}_{ij} = t_{ij1} + it_{ij2}$, where t_{ij1} and t_{ij2} are real, we have $|\tilde{t}_{ij}|^{2} = t_{ij1}^{2} + t_{ij2}^{2}$. Hence

$$\int_{\tilde{t}_{ij}} e^{-|\tilde{t}_{ij}|^{2}}d\tilde{t}_{ij} = \int_{-\infty}^{\infty}\int_{-\infty}^{\infty} e^{-t_{ij1}^{2}-t_{ij2}^{2}}dt_{ij1} \wedge dt_{ij2} = \pi.$$

But $\prod_{i>j}\pi = \pi^{\frac{p(p-1)}{2}}$. Taking all products together we have

$$\pi^{\frac{p(p-1)}{2}}\Gamma(\alpha)\Gamma(\alpha-1)\ldots\Gamma(\alpha-p+1) = \tilde{\Gamma}_{p}(\alpha), \quad \Re(\alpha) > p-1.$$

This establishes the result.

From here, we can define matrix-variate gamma and beta densities for the complex domain.

Definition 2.6.2. *A complex matrix-variate gamma density with parameters* α, B. *For* $B = B^{*} > O$ *a* $p \times p$ *hermitian positive definite constant matrix,* $\tilde{X}$ *a* $p \times p$ *hermitian positive definite matrix of distinct complex variables as elements, except for the property* of hermitian positive definiteness, then the complex matrix-variate gamma density is defined as

$$f(\tilde{X}) = \begin{cases} \dfrac{|\det(B)|^{\alpha}}{\tilde{\Gamma}_p(\alpha)} |\det(\tilde{X})|^{\alpha-p} e^{-\text{tr}(\tilde{X})}, & \tilde{X} = \tilde{X}^* > O, \\ \qquad B > O, \ \Re(\alpha) > p - 1, \\ 0 \quad \text{elsewhere.} \end{cases} \qquad (2.6.8)$$

Definition 2.6.3. *Complex matrix-variate beta function*: It is denoted by $\tilde{B}_P(\alpha, \beta)$ and is defined as

$$\tilde{B}_p(\alpha, \beta) = \frac{\tilde{\Gamma}_p(\alpha)\tilde{\Gamma}_p(\beta)}{\tilde{\Gamma}_p(\alpha + \beta)}, \Re(\alpha) > p - 1, \Re(\beta) > p - 1. \qquad (2.6.9)$$

Definition 2.6.4. *Complex matrix-variate type-1 beta density*: The following two densities are type-1 beta densities in the complex case where in both cases the conditions $\Re(\alpha) > p - 1, \Re(\beta) > p - 1$ are needed and the non-zero part is over $O < \tilde{X} < I$ and it is zero outside. Only the non-zero parts are given in all the definitions to follow.

$$f_1(\tilde{X}) = \frac{\tilde{\Gamma}_p(\alpha + \beta)}{\tilde{\Gamma}_p(\alpha)\tilde{\Gamma}_p(\beta)} |\det(\tilde{X})|^{\alpha-p} |\det(I - \tilde{X})|^{\beta-p}, O < \tilde{X} < I,$$

$$(2.6.10)$$

$$f_2(\tilde{X}) = \frac{1}{\tilde{B}_p(\alpha, \beta)} |\det(\tilde{X})|^{\beta-p} |\det(I - \tilde{X})|^{\alpha-p}, O < \tilde{X} < I.$$

$$(2.6.11)$$

Definition 2.6.5. *Complex matrix-variate type-2 beta densities*: The following two forms are complex matrix-variate type-2 beta densities where $\Re(\alpha) > p - 1, \Re(\beta) > p - 1$ and $\tilde{X} = \tilde{X}^* > O$:

$$f_3(\tilde{X}) = \frac{1}{\tilde{B}_p(\alpha, \beta)} |\det(\tilde{X})|^{\alpha-p} |\det(I + \tilde{X})|^{-(\alpha+\beta)}, \quad \tilde{X} > O,$$

$$(2.6.12)$$

$$f_4(\tilde{X}) = \frac{1}{\tilde{B}_p(\alpha, \beta)} |\det(\tilde{X})|^{\beta-p} |\det(I - \tilde{X})|^{-(\alpha+\beta)}, \quad \tilde{X} > O.$$

$$(2.6.13)$$

100   *Matrix Methods and Fractional Calculus*

Definition 2.6.6 *A complex p-variate Gaussian density:* Let $\tilde{X}$ be a $p \times 1$ complex vector of distinct p complex variables. Let $\tilde{\mu}$ be a $p \times 1$ constant vector and V be $p \times p$ hermitian positive definite matrix, $V = V^* > O$. Then the p-variate complex Gaussian density is defined as follows:

$$g(\tilde{X}) = \frac{1}{(2\pi)^p |\det(V)|^p} e^{-(\tilde{X}-\tilde{\mu})^* V^{-1}(\tilde{X}-\tilde{\mu})}. \tag{2.6.14}$$

Definition 2.6.7. *A complex matrix-variate Gaussian density:* Let $\tilde{X}$ be an $m \times n$ matrix of distinct complex elements. Let M be an $m \times n$ constant matrix. Let A be $m \times m$ and B be $n \times n$ constant hermitian positive definite matrices. Then the following density is the complex matrix-variate Gaussian density:

$$g_1(\tilde{X}) = \frac{|\det(A)|^n |\det(B)|^m}{(2\pi)^{mn}} e^{-\text{tr}[A(\tilde{X}-M)^* B(\tilde{X}-M)]}. \tag{2.6.15}$$

Result 2.6.6. *Let $\tilde{X}$ be a $p \times p$ non-singular matrix of distinct complex variables. Then, ignoring the sign,*

$$\tilde{Y} = \tilde{X}^{-1} \Rightarrow d\tilde{Y} = \begin{cases} |\det(\tilde{X}\tilde{X}^*)|^{-2p} d\tilde{X} & \textit{for a general } \tilde{X} \\ |\det(\tilde{X}\tilde{X}^*)|^{-p} d\tilde{X} & \textit{for } \tilde{X} = \tilde{X}^* \textit{ or } \tilde{X} = -\tilde{X}^*. \end{cases} \tag{2.6.16}$$

Result 2.6.7. *Let $\tilde{T}$ be a lower triangular matrix of distinct complex variables with the diagonal elements real and positive. Let $\tilde{U}$ be a unitary matrix, $\tilde{U}\tilde{U}^* = I, \tilde{U}^*\tilde{U} = I$. Let $\tilde{X} = \tilde{T}\tilde{U}$. Then*

$$\tilde{X} = \tilde{T}\tilde{U} \Rightarrow d\tilde{X} = \left\{ \prod_{j=1}^p t_{jj}^{2(p-j)+1} \right\} d\tilde{T} \wedge d\tilde{G}, \tag{2.6.17}$$

where $(d\tilde{G}) = \tilde{U}(d\tilde{U}^*)$.

Result 2.6.8. *Let $\tilde{X}$ be $p \times p$ hermitian matrix of distinct complex variables with distinct eigenvalues $\lambda_1 > \lambda_2 > \cdots > \lambda_p$. Let $\tilde{U}$ be a $p \times p$ unique unitary matrix. Let $\tilde{U}^*\tilde{X}\tilde{U} = D = \text{diag}(\lambda_1, \ldots, \lambda_p)$ or*

$\tilde{X} = \tilde{U}D\tilde{U}^*$. *Then*

$$\tilde{X} = \tilde{U}D\tilde{U}^* \Rightarrow d\tilde{X} = \left\{ \prod_{j>k} |\lambda_k - \lambda_j|^2 \right\} dD \wedge d\tilde{G}, \qquad (2.6.18)$$

where $(d\tilde{G}) = \tilde{U}(d\tilde{U}^*)$.

Result 2.6.9. *Let* $\tilde{X}$ *be* $p \times n, n \geq p$ *matrix of distinct complex variables. Let* $\tilde{X} = \tilde{T}\tilde{U}$ *where* $\tilde{T}$ *is* $p \times p$ *lower triangular matrix of distinct complex variables with the diagonal elements real and positive. Let* $\tilde{U}$ *be a* $p \times n$ *semiunitary matrix such that* $\tilde{U}\tilde{U}^* = I_p$. *Then*

$$\tilde{X} = \tilde{T}\tilde{U} \Rightarrow d\tilde{X} = \left\{ \prod_{j=1}^{p} t_{jj}^{2(n-j)} \right\} d\tilde{T} \wedge d\tilde{G}, \qquad (2.6.19)$$

where $(d\tilde{G}) = \tilde{U}(d\tilde{U}^*)$.

Result 2.6.10. *Let* $\tilde{U}$ *be a* $p \times n$ *semiunitary matrix,* $\tilde{U}\tilde{U}^* = I_p$. *Let* dG *be the wedge product of the differentials in* $(d\tilde{G}) = \tilde{U}(d\tilde{U}^*)$. *Then*

$$\int_{\tilde{U}} d\tilde{G} = \begin{cases} \dfrac{2^p \pi^{np}}{\tilde{\Gamma}_p(n)} & \text{for a general } \tilde{U}, n \geq p, \\[3mm] \dfrac{2^p \pi^{p^2}}{\tilde{\Gamma}_p(p)} & \text{for a general } \tilde{U}, n = p, \\[3mm] \dfrac{2^p \pi^{p(n-1)}}{\tilde{\Gamma}_p(n)} & \text{for } \tilde{U} \text{ with } u_{jj}\text{'s real.} \end{cases} \qquad (2.6.20)$$

Result 2.6.11. *Let* $\tilde{X}$ *be* $p \times n, n \geq p$ *matrix of distinct complex variables. Let* $\tilde{S} = \tilde{X}\tilde{X}^*$. *Then*

$$d\tilde{X} = \frac{\pi^{np}}{\tilde{\Gamma}_p(n)} |\det(\tilde{S})|^{n-p} d\tilde{S}. \qquad (2.6.21)$$

This Result 2.6.13 is a very important result when we want to go from a rectangular matrix to a square matrix.

Exercise 2.6

[Work out all the following exercises from first principles. Do not use any result given in Section 2.6. Use only the definitions]

2.6.1. Let $\tilde{y} = a\tilde{x} + b, \tilde{x} = x_1 + iy_1, a = 3 - 2i, i = \sqrt{-1}, b = 4 + 8i$. Obtain (1): $d\tilde{y}$ in terms of $d\tilde{x}$, (2): $d\tilde{y}^*$ in terms of $d\tilde{y}$, (3): $d\tilde{y}^*$ in terms of $d\tilde{x}^*$, (4): $d\tilde{y}^*$ in terms of $d\tilde{x}$.

2.6.2. Let $\tilde{Y} = A\tilde{X}, A = \begin{bmatrix} 2+i & 1-i \\ 1+2i & 3 \end{bmatrix}, \tilde{X} = \begin{bmatrix} \tilde{x}_{11} \\ \tilde{x}_{21} \end{bmatrix}, \tilde{Y} = \begin{bmatrix} \tilde{y}_{11} \\ \tilde{y}_{21} \end{bmatrix}$. Write $d\tilde{Y}$ in terms of $d\tilde{X}$.

2.6.3. Repeat Exercise 2.6.2 if A is the same, $\tilde{X} = \begin{bmatrix} \tilde{x}_{11} & \tilde{x}_{12} \\ \tilde{x}_{21} & \tilde{x}_{22} \end{bmatrix}, \tilde{Y} = \begin{bmatrix} \tilde{y}_{11} & \tilde{y}_{12} \\ \tilde{y}_{21} & \tilde{y}_{22} \end{bmatrix}$.

2.6.4. Let $\tilde{Y} = A\tilde{X}B$ where $A = \begin{bmatrix} 1-i & 2+3i \\ 2-i & 3+4i \end{bmatrix}, B = \begin{bmatrix} 2i & 1-i \\ 2+i & 1+2i \end{bmatrix}$. Let $\tilde{X}$ and $\tilde{Y}$ be 2×2 matrices of distinct complex variables each. Evaluate $d\tilde{Y}$ in terms of $d\tilde{X}$.

2.6.5. Write two examples each of a 2×2 and 3×3 hermitian matrices (not diagonal types).

2.6.6. Write down two examples each of a 2×2 and 3×3 hermitian positive definite matrices (not diagonal types).

2.6.7. Let $\tilde{Y} = A\tilde{X}A^*$ where $\tilde{X}$ and $\tilde{Y}$ are hermitian matrices and (1): $A = \begin{bmatrix} 2 & 1+i \\ 3-i & 5 \end{bmatrix}$, (2): $A = \begin{bmatrix} 2 & -i \\ 1+i & 5 \end{bmatrix}$. Evaluate $d\tilde{Y}$ in terms of $d\tilde{X}$ in each case.

2.6.8. Evaluate the determinant $\det(A)$ and absolute value of the determinant $|\det(A)|$ for each case of A where

$$\text{(1): } A = \begin{bmatrix} 2+i & 1+i \\ 1+2i & 3+2i \end{bmatrix}, \quad \text{(2): } A = \begin{bmatrix} 2+i & 1+i & 0 \\ 1+2i & 3+2i & 2 \\ 0 & 3 & 4+i \end{bmatrix}.$$

2.6.9. Evaluate A^{-1}, if it exists, for each case in Exercise 2.6.8.

2.6.10. Let $\tilde{X}$ be a 2×2 matrix of distinct complex variables. Let $\det(\tilde{X}) \neq 0$. Let $\tilde{Y} = \tilde{X}^{-1}$. Evaluate $d\tilde{Y}$ in terms of $d\tilde{X}$.

2.6.11. Repeat Exercise 2.6.10 if $\tilde{X}$ is (i): hermitian, (ii): skew hermitian, from first principles.

2.6.12. Let $\tilde{X}$ be a 2×2 hermitian positive definite matrix. Let $\tilde{T}$ be a lower triangular matrix with real and positive diagonal elements. (i): Show that the transformation $\tilde{X} = \tilde{T}\tilde{T}^*$ is one-to-one. (ii): Evaluate $\mathrm{d}\tilde{X}$ in terms of $\mathrm{d}\tilde{T}$ from first principles.

2.6.13. Repeat Exercise 2.6.12 if $\tilde{X}$ is a 3×3 hermitian positive definite matrix.

2.6.14. Let $\tilde{X}$ be a 2×2 hermitian positive definite matrix. Evaluate $\int_{\tilde{X}>O} \mathrm{e}^{-\mathrm{tr}(\tilde{X})} \mathrm{d}\tilde{X}$ by direct integration without using the result in Section 2.6 and then verify with the result in Section 2.6.

2.6.15. Let $\tilde{X}$ be a 2×2 hermitian positive definite matrix with all its eigenvalues in the open interval $(0, 1)$. Evaluate the following integrals:

(i): $\int_{\tilde{X}>O} \mathrm{d}\tilde{X}$; (ii): $\int_{\tilde{X}>O} |\det(\tilde{X})| \mathrm{d}\tilde{X}$; (iii): $\int_{\tilde{X}>O} |\det(I-\tilde{X})| \mathrm{d}\tilde{X}$

by direct integration and then verify the answers by using the results in Section 2.6.

2.6.16. Construct two examples each of (1): a 2×2, (2): a 3×3 unitary matrix. (Identity matrix is not acceptable as an answer.)

2.6.17. By using your unitary matrices in Exercise 2.6.16 construct two examples each of (1): hermitian positive definite matrix, (2): hermitian negative definite matrix, (3): hermitian indefinite matrix.

2.6.18. Let $\tilde{X}$ be a $p \times p$ hermitian positive definite matrix of distinct complex variables. Let B be a $p \times p$ hermitian positive definite constant matrix. Let $f(\tilde{X}) = c\mathrm{e}^{-\mathrm{tr}(B\tilde{X})}$. Evaluate c if $f(\tilde{X})$ is a density. [A non-negative function with total integral unity is a density function.]

2.6.19. If $f(\tilde{X}) = c_1 \mathrm{e}^{-\mathrm{tr}(AXBX^*)}$ is a density, where A is $m \times m$, B is $n \times n$ constant positive definite matrices and $\tilde{X}$ is an $m \times n$ matrix of distinct complex elements, then evaluate c_1.

2.6.20. Write down $\tilde{\Gamma}_p(\alpha)$ explicitly for (1): $p = 1$, (2): $p = 2$, (3): $p = 3$ for a general α as well as for $\alpha = 3.5$.

Acknowledgments

The author would like to thank the Department of Science and Technology, Government of India, for the financial assistance for this work under project number SR/S4/MS:287/05 and the Centre for Mathematical and Statistical Sciences for the facilities.

Bibliography

[1]   A.M. Mathai, Some properties of Mittag-Leffler functions and matrix-variate analogues: A statistical perspective, *Fract. Calc. Appl. Anal.* **13**(1) (2010), 113–132.

[2]   A.M. Mathai, Fractional integrals in the complex matrix-variate case, *Linear Algebra Appl.* **439**(2013), 2901–2913.

[3]   A.M. Mathai, *Jacobians of Matrix Transformations and Functions of Matrix Argument*, World Scientific Publishing, New York, 1997.

Chapter 3

Fractional Calculus and Special Functions*

3.1. Introduction

Fractional calculus is the field of mathematical analysis which deals with the investigation and applications of integrals and derivatives of arbitrary order. The term *fractional* is a misnomer, but it is retained following the prevailing use. For details on the historical development of the fractional calculus we refer the interested reader to Ross' bibliography in [138] and to the historical notes generally available in any subsequent text on Fractional Calculus.

In recent years considerable interest in fractional calculus has been stimulated by the applications that it finds in different fields of science, including numerical analysis, economics and finance, engineering, physics, biology, etc.

We use the standard notations $\mathbb{N}$, $\mathbf{Z}$, $\mathbb{R}$, $\mathbb{C}$ to denote the sets of natural, integer, real and complex numbers, respectively; furthermore, $\mathbb{R}^+$ and $\mathbb{R}_0^+$ denote the sets of positive real numbers and of non-negative real numbers, respectively.

For the continuous approach to Fractional Calculus we distinguish three types: the *Riemann–Liouville fractional calculus*, the *Liouville–Weyl fractional calculus*, the *Riesz–Feller fractional calculus*, which, concerning three different types of integral operators acting in $\mathbb{R}$, are of major interest for us. We shall devote the next three sections (Sections 3.2–3.4) to the above kinds of fractional calculus,

---

*This chapter is based on lectures of Professor Dr Francesco Mainardi of the University of Bologna, Italy.

respectively. In these sections the independent real variable will be denoted by x. We devote Section 3.5 to the Riemann–Liouville fractional calculus and its variant, known as Caputo fractional calculus, by taking the integral operators acting on the variable $t \in \mathbb{R}^+$.

Sections 3.5 and 3.6 are devoted, respectively, to the special (higher transcendental) functions of the Mittag-Leffler and Wright type, which play a fundamental role in our applications of the fractional calculus.

Let us remark that, wanting our lectures to be accessible to various kinds of people working in applications (e.g., physicists, chemists, theoretical biologists, economists, engineers) here we, deliberately and consciously as far as possible, avoid the language of functional analysis. We thus use vague phrases like "for a sufficiently well-behaved function" instead of constructing a stage of precisely defined spaces of admissible functions. We pay particular attention to the techniques of integral transforms: we limit ourselves to Fourier, and Laplace transforms for which notations and main properties will be recalled as soon as they become necessary. We make formal use of generalized functions related to the Dirac "delta function" in the typical way suitable for applications in physics and engineering, without adopting the language of distributions. We kindly ask specialists of these fields of pure mathematics to forgive us. Our notes are written in a way that makes it easy to fill in details of precision which in their opinion might be lacking.

The bibliography contains a remarkably large number of references to articles and books not all mentioned in the text, since they have attracted the author's attention over the last decades and cover topics more or less related to these lectures. The interested reader could hopefully take advantage of this bibliography, even if not exhaustive, for enlarging and improving the scope of these lectures.

3.2. Riemann–Liouville Fractional Calculus

We now present an introductory survey of the Riemann–Liouville (RL) fractional calculus. As it is customary, let us take as our starting point for the development of this approach the repeated integral

$$I_{a+}^n \, \phi(x) := \int_a^x \int_a^{x_{n-1}} \dots \int_a^{x_1} \phi(x_0) \, \mathrm{d}x_0 \dots \mathrm{d}x_{n-1},$$

$$a \leq x < b, \ n \in \mathbb{N}, \tag{3.2.1}$$

where $a > -\infty$ and $b \leq +\infty$. The function $\phi(x)$ is assumed to be well behaved; for this it suffices that $\phi(x)$ is locally integrable in the interval $[a, b)$, meaning in particular that a possible singular behavior at $x = a$ does not destroy integrability. It is well known that the above formula provides an n-fold primitive $\phi_n(x)$ of $\phi(x)$, precisely that primitive which vanishes at $x = a$ jointly with its derivatives of order $1, 2, \dots, n - 1$.

We can rewrite this n-fold repeated integral by a convolution-type formula (often attributed to Cauchy) as

$$I_{a+}^n \, \phi(x) = \frac{1}{(n-1)!} \int_a^x (x - \xi)^{n-1} \phi(\xi) \, \mathrm{d}\xi, \quad a \leq x < b. \tag{3.2.2}$$

In a natural way we are now led to extend the formula (3.2.2) from positive integer values of the index n to arbitrary positive values α, thereby using the relation $(n - 1)! = \Gamma(n)$. So, using the Gamma function, we define the *fractional integral of order* α as

$$I_{a+}^\alpha \, \phi(x) := \frac{1}{\Gamma(\alpha)} \int_a^x (x - \xi)^{\alpha-1} \phi(\xi) \, \mathrm{d}\xi, \quad a < x < b, \ \alpha > 0.$$

$$\tag{3.2.3}$$

We remark that the values $I_{a+}^n \, \phi(x)$ with $n \in \mathbb{N}$ are always finite for $a \leq x < b$, but the values $I_{a+}^\alpha \, \phi(x)$ for $\alpha > 0$ are assumed to be finite for $a < x < b$, whereas, as we shall see later, it may happen that the limit (if it exists) of $I_{a+}^\alpha \phi(x)$ for $x \to a^+$, that we denote by $I_{a+}^\alpha \, \phi(a^+)$, is infinite.

Without loss of generality, it may be convenient to set $a = 0$. We agree to refer to the fractional integrals I_{0+}^α as to the

Riemann–Liouville fractional integrals, honoring both the authors who first treated similar integrals.

For I_{0+} we use the special and simplified notation J^α in agreement with the notation introduced by Gorenflo and Vessella [59] and then followed in all our papers in the subject. We shall return to the fractional integrals J^α in Section 3.5 providing a sufficiently exhaustive treatment of the related fractional calculus.[a]

A dual form of the integrals (3.2.2) is

$$I_{b-}^n \phi(x) = \frac{1}{(n-1)!} \int_x^b (\xi - x)^{n-1} \phi(\xi)\, d\xi, \quad a < x \le b, \quad (3.2.4)$$

where we assume $\phi(x)$ to be sufficiently well behaved in $-\infty \le a < x < b < +\infty$. Now it suffices that $\phi(x)$ is locally integrable in $(a, b]$.

Extending (3.2.4) from the positive integers n to $\alpha > 0$ we obtain the dual form of the fractional integral (3.2.3), i.e.,

$$I_{b-}^\alpha \phi(x) := \frac{1}{\Gamma(\alpha)} \int_x^b (\xi - x)^{\alpha-1} \phi(\xi)\, d\xi, \quad a < x \le b, \quad \alpha > 0.$$
$$(3.2.5)$$

Now it may happen that the limit (if it exists) of $I_{b-}^\alpha \phi(x)$ for $x \to b^-$, that we denote by $I_{b-}^\alpha \phi(b^-)$, is infinite.

We refer to the fractional integrals I_{a+}^α and I_{b-}^α as *progressive* or right-handed and *regressive* or left-handed, respectively.

Let us point out the fundamental property of the fractional integrals, namely the additive index law (*semi-group property*) according to which

$$I_{a+}^\alpha I_{a+}^\beta = I_{a+}^{\alpha+\beta}, \quad I_{b-}^\alpha I_{b-}^\beta = I_{b-}^{\alpha+\beta}, \quad \alpha, \beta \ge 0, \quad (3.2.6)$$

---

[a]Historically, fractional integrals of type (3.2.3) were first investigated in papers by Abel [1, 2] and by Riemann [136]. In fact, Abel, when he introduced his integral equation, named after him, to treat the problem of the tautochrone, was able to find the solution inverting a fractional integral of type (3.2.3). The contribution by Riemann is supposed to be independent from Abel and inspired by previous works by Liouville (see later): it was written in January 1847 when he was still a student, but published in 1876, 10 years after his death.

where, for complementation, we have defined $I_{a+}^0 = I_{b-}^0 := \mathbb{I}$ (Identity operator) which means $I_{a+}^0 \, \phi(x) = I_{b-}^0 \, \phi(x) = \phi(x)$. The proof of (3.2.6) is based on Dirichlet's formula concerning the change of the order of integration and the use of the Beta function in terms of Gamma function.

We note that the fractional integrals (3.2.3) and (3.2.5) contain a weakly singular kernel only when the order is less than one.

We can now introduce the concept of fractional derivative based on the fundamental property of the common derivative of integer order n

$$D^n \, \phi(x) = \frac{d^n}{dx^n} \, \phi(x) = \phi^{(n)}(x), \quad a < x < b$$

to be the *left inverse operator* of the progressive repeated integral of the same order n, $I_{a+}^n \, \phi(x)$. In fact, it is straightforward to recognize that the derivative of any integer order $n = 0, 1, 2, \ldots$ satisfies the following composition rules with respect to the repeated integrals of the same order n, $I_{a+}^n \, \phi(x)$ and $I_{b-}^n \, \phi(x)$:

$$\begin{cases} D^n \, I_{a+}^n \, \phi(x) = \phi(x), \\ I_{a+}^n \, D^n \, \phi(x) = \phi(x) - \displaystyle\sum_{k=0}^{n-1} \frac{\phi^{(k)}(a^+)}{k!} \, (x-a)^k, \end{cases} \quad a < x < b, \qquad (3.2.7)$$

$$\begin{cases} D^n \, I_{b-}^n \, \phi(x) = (-1)^n \phi(x), \\ I_{b-}^n \, D^n \, \phi(x) \\ \quad = (-1)^n \left\{ \phi(x) - \displaystyle\sum_{k=0}^{n-1} (-1)^k \, \frac{\phi^{(k)}(b^-)}{k!} \, (b-x)^k \right\}, \end{cases} \quad a < x < b.$$

$$(3.2.8)$$

In view of the above properties we can define the progressive/regressive fractional derivatives of order $\alpha > 0$ provided that we first introduce the positive integer m such that $m - 1 < \alpha \leq m$. Then we define the *progressive/regressive fractional derivative of order* α, $D_{a+}^\alpha / D_{b-}^\alpha$, as the *left inverse* of the corresponding fractional integral $I_{a+}^\alpha / I_{b-}^\alpha$. As a consequence of the previous formulas (3.2.2)–(3.2.8) the *fractional derivatives of order* α turn out to be defined as follows:

$$\begin{cases} D_{a+}^\alpha \, \phi(x) := D^m \, I_{a+}^{m-\alpha} \, \phi(x), \\ D_{b-}^\alpha \, \phi(x) := (-1)^m \, D^m \, I_{b-}^{m-\alpha} \, \phi(x), \end{cases} \quad a < x < b, \; m - 1 < \alpha \leq m.$$

$$(3.2.9)$$

In fact, for the progressive/regressive derivatives we get

$$\begin{cases} D_{a+}^\alpha \, I_{a+}^\alpha = D^m \, I_{a+}^{m-\alpha} \, I_{a+}^\alpha = D^m \, I_{a+}^m = \mathbb{I}, \\ D_{b-}^\alpha \, I_{b-}^\alpha = (-1)^m \, D^m \, I_{b-}^{m-\alpha} \, I_{b-}^\alpha = (-1)^m \, D^m \, I_{b-}^m = \mathbb{I}. \end{cases}$$

$$(3.2.10)$$

For complementation we also define $D_{a+}^0 = D_{b-}^0 := \mathbb{I}$ (Identity operator).

When the order is not integer ($m - 1 < \alpha < m$) the definitions (3.2.9) with (3.2.3), (3.2.5) yield the explicit expressions

$$D_{a+}^\alpha \, \phi(x) = \frac{1}{\Gamma(m - \alpha)} \frac{\mathrm{d}^m}{\mathrm{d}x^m} \int_a^x (x - \xi)^{m-\alpha-1} \, \phi(\xi) \, \mathrm{d}\xi, \quad a < x < b,$$

$$(3.2.11)$$

$$D_{b-}^\alpha \, \phi(x) = \frac{(-1)^m}{\Gamma(m - \alpha)} \frac{\mathrm{d}^m}{\mathrm{d}x^m} \int_x^b (\xi - x)^{m-\alpha-1} \, \phi(\xi) \, \mathrm{d}\xi, \quad a < x < b.$$

$$(3.2.12)$$

We stress the fact that the "proper" fractional derivatives (namely when α is non-integer) are *non-local operators* being expressed by ordinary derivatives of convolution-type integrals with a weakly singular kernel, as it is evident from (3.2.11)–(3.2.12). Furthermore they do not obey necessarily the analogue of the "semi-group" property of the fractional integrals: in this respect the initial point $a \in \mathbb{R}$ (or the ending point $b \in \mathbb{R}$) plays a "disturbing" role. We also note that when the order α tends to both the endpoints of the interval $(m-1, m)$ we recover the common derivatives of integer order (unless of the sign for the regressive derivative).

3.3. Liouville–Weyl Fractional Calculus

We can also define the fractional integrals over unbounded intervals and, as left inverses, the corresponding fractional derivatives. If the function $\phi(x)$ is locally integrable in $-\infty < x < b \leq +\infty$, and behaves for $x \to -\infty$ in such a way that the following integral exists in an appropriate sense, we can define the *Liouville fractional integral of order* α as

$$I_+^\alpha \, \phi(x) := \frac{1}{\Gamma(\alpha)} \int_{-\infty}^x (x - \xi)^{\alpha-1} \, \phi(\xi) \, d\xi, \quad -\infty < x < b, \ \alpha > 0.$$
(3.3.1)

Analogously, if the function $\phi(x)$ is locally integrable in $-\infty \le a < x < +\infty$, and behaves well enough for $x \to +\infty$, we define the *Weyl fractional integral of order* α as

$$I_-^\alpha \, \phi(x) := \frac{1}{\Gamma(\alpha)} \int_x^{+\infty} (\xi - x)^{\alpha-1} \, \phi(\xi) \, d\xi, \quad a < x < +\infty, \ \alpha > 0.$$
(3.3.2)

Note the kernel $(\xi - x)^{\alpha-1}$ for (3.3.2). The names of Liouville and Weyl are here adopted for the fractional integrals (3.3.1), (3.3.2), respectively, following a standard terminology, see, e.g., [19], based on historical reasons.

We note that a sufficient condition for the integrals entering $I_\pm^\alpha$ in (3.3.1)–(3.3.2) to converge is that

$$\phi(x) = O(|x|^{-\alpha-\epsilon}), \quad \epsilon > 0, \quad x \to \mp\infty,$$

respectively. Integrable functions satisfying these properties are sometimes referred to as functions of Liouville class (for $x \to -\infty$), and of Weyl class (for $x \to +\infty$), respectively, see [123]. For example, power functions $|x|^{-\delta}$ with $\delta > \alpha > 0$ and exponential functions e^{cx} with $c > 0$ are of Liouville class (for $x \to -\infty$).[b] For these functions we obtain

$$\begin{cases} I_+^\alpha \, |x|^{-\delta} = \dfrac{\Gamma(\delta - \alpha)}{\Gamma(\delta)} \, |x|^{-\delta+\alpha}, \\[4mm] D_+^\alpha \, |x|^{-\delta} = \dfrac{\Gamma(\delta + \alpha)}{\Gamma(\delta)} \, |x|^{-\delta-\alpha}, \end{cases} \quad \delta > \alpha > 0, \ x < 0, \quad (3.3.3)$$

---

[b]In fact, Liouville considered in a series of papers from 1832 to 1837 the integrals of progressive type (3.3.1), see, e.g., [82–84]. On the other hand, Weyl [159] arrived at the regressive integrals of type (3.3.2) indirectly by defining fractional integrals suitable for periodic functions.

and

$$\begin{cases} I_+^\alpha \, e^{cx} = c^{-\alpha} \, e^{cx}, \\ D_+^\alpha \, e^{cx} = c^\alpha \, e^{cx}, \end{cases} \quad c > 0, \quad x \in \mathbb{R}. \qquad (3.3.4)$$

Also for the Liouville and Weyl fractional integrals we can state the corresponding *semigroup property*

$$I_+^\alpha \, I_+^\beta = I_+^{\alpha+\beta}, \quad I_-^\alpha \, I_-^\beta = I_-^{\alpha+\beta}, \quad \alpha, \, \beta \geq 0, \qquad (3.3.5)$$

where, for complementation, we have defined $I_+^0 = I_-^0 := \mathbb{I}$ (Identity operator). For more details on Liouville–Weyl fractional integrals we refer to [122, 123, 143].

For the definition of the *Liouville–Weyl fractional derivatives of order* α we follow the scheme adopted in the previous section for bounded intervals. Having introduced the positive integer m so that $m - 1 < \alpha \leq m$ we define

$$\begin{cases} D_+^\alpha \, \phi(x) := D^m \, I_+^{m-\alpha} \, \phi(x), & -\infty < x < b, \\ D_-^\alpha \, \phi(x) := (-1)^m \, D^m \, I_-^{m-\alpha} \, \phi(x), & a < x < +\infty, \end{cases} \quad (m-1 < \alpha \leq m),$$

$$(3.3.6)$$

with $D_+^0 = D_-^0 = \mathbb{I}$. In fact we easily recognize using (3.3.5)–(3.3.6) the fundamental property

$$D_+^\alpha \, I_+^\alpha = \mathbb{I} = (-1)^m D_-^\alpha \, I_-^\alpha. \qquad (3.3.7)$$

The explicit expressions for the "proper" Liouville and Weyl fractional derivatives $(m - 1 < \alpha < m)$ read

$$D_+^\alpha \, \phi(x) = \frac{1}{\Gamma(m-\alpha)} \frac{\mathrm{d}^m}{\mathrm{d}x^m} \int_{-\infty}^x (x-\xi)^{m-\alpha-1} \phi(\xi) \, \mathrm{d}\xi, \quad x \in \mathbb{R},$$

$$(3.3.8)$$

$$D_-^\alpha \, \phi(x) = \frac{(-1)^m}{\Gamma(m-\alpha)} \frac{\mathrm{d}^m}{\mathrm{d}x^m} \int_x^{+\infty} (\xi-x)^{m-\alpha-1} \phi(\xi) \, \mathrm{d}\xi, \quad x \in \mathbb{R}.$$

$$(3.3.9)$$

Because of the unbounded intervals of integration, fractional integrals and derivatives of Liouville and Weyl type can be (successfully) handled via Fourier transform and related theory of pseudo-differential operators, that, as we shall see, simplifies their treatment. For this purpose, let us now recall our notations and the relevant results concerning the Fourier transform.

Let

$$\widehat{\phi}(\kappa) = \mathcal{F}\{\phi(x); \kappa\} = \int_{-\infty}^{+\infty} e^{+i\kappa x}\,\phi(x)\,dx, \quad \kappa \in \mathbb{R}, \qquad (3.3.10)$$

be the Fourier transform of a sufficiently well-behaved function $\phi(x)$, and let

$$\phi(x) = \mathcal{F}^{-1}\{\widehat{\phi}(\kappa); x\} = \frac{1}{2\pi} \int_{-\infty}^{+\infty} e^{-i\kappa x}\,\widehat{\phi}(\kappa)\,d\kappa, \quad x \in \mathbb{R},$$
$$(3.3.11)$$

denote the inverse Fourier transform.

In this framework we also consider the class of pseudo-differential operators of which the ordinary repeated integrals and derivatives are special cases.[c] A pseudo-differential operator A, acting with respect to the variable $x \in \mathbb{R}$, is defined through its Fourier representation, namely

$$\int_{-\infty}^{+\infty} e^{i\kappa x}\,A\,\phi(x)\,dx = \widehat{A}(\kappa)\,\widehat{\phi}(\kappa), \qquad (3.3.12)$$

where $\widehat{A}(\kappa)$ is referred to as symbol of A. An often applicable practical rule is

$$\widehat{A}(\kappa) = (A\,e^{-i\kappa x})\,e^{+i\kappa x}, \quad \kappa \in \mathbb{R}. \qquad (3.3.13)$$

If B is another pseudo-differential operator, then we have

$$\widehat{A\,B}(\kappa) = \widehat{A}(\kappa)\,\widehat{B}(\kappa). \qquad (3.3.14)$$

---

[c]In the ordinary theory of the Fourier transform the integral in (3.3.10) is assumed to be a "Lebesgue integral" whereas the one in (3.3.11) can be the "principal value" of a "generalized integral". In fact, $\phi(x) \in L_1(\mathbb{R})$, necessary for writing (3.3.10), is not sufficient to ensure $\widehat{\phi}(\kappa) \in L_1(\mathbb{R})$. However, we allow for an extended use of the Fourier transform which includes Dirac-type generalized functions: then the above integrals must be properly interpreted in the framework of the theory of distributions.

For the sake of convenience we adopt the notation $\overset{\mathcal{F}}{\leftrightarrow}$ to denote the juxtaposition of a function with its Fourier transform and that of a pseudo-differential operator with its symbol, namely

$$\phi(x) \overset{\mathcal{F}}{\leftrightarrow} \widehat{\phi}(\kappa), \quad A \overset{\mathcal{F}}{\leftrightarrow} \widehat{A}. \tag{3.3.15}$$

We now consider the pseudo-differential operators represented by the Liouville–Weyl fractional integrals and derivatives. Of course we assume that the integrals entering their definitions are in a proper sense, in order to ensure that the resulting functions of x can be Fourier transformable in ordinary or generalized sense.

The symbols of the fractional Liouville–Weyl integrals and derivatives can be easily derived according to

$$\begin{cases} \widehat{I_\pm^\alpha} = (\mp i\kappa)^{-\alpha} = |\kappa|^{-\alpha}\, e^{\pm i\,(\operatorname{sign}\kappa)\,\alpha\pi/2}, \\[2mm] \widehat{D_\pm^\alpha} = (\mp i\kappa)^{+\alpha} = |\kappa|^{+\alpha}\, e^{\mp i\,(\operatorname{sign}\kappa)\,\alpha\pi/2}. \end{cases} \tag{3.3.16}$$

Based on a former idea by Marchaud, see, e.g., [64, 112, 143], we now point out purely integral expressions for $D_\pm^\alpha$ which are alternative to the integro-differential expressions (3.3.8) and (3.3.9).

We limit ourselves to the case $0 < \alpha < 1$. Let us first consider from Eq. (3.3.8) the progressive derivative

$$D_+^\alpha = \frac{\mathrm{d}}{\mathrm{d}x}\, I_+^{1-\alpha}, \quad 0 < \alpha < 1. \tag{3.3.17}$$

We have, see [64],

$$D_+^\alpha\, \phi(x) = \frac{\mathrm{d}}{\mathrm{d}x}\, I_+^{1-\alpha}\, \phi(x)$$

$$= \frac{1}{\Gamma(1-\alpha)}\, \frac{\mathrm{d}}{\mathrm{d}x} \int_{-\infty}^{x} (x-\xi)^{-\alpha}\, \phi(\xi)\, \mathrm{d}\xi$$

$$= \frac{1}{\Gamma(1-\alpha)}\, \frac{\mathrm{d}}{\mathrm{d}x} \int_{0}^{\infty} \xi^{-\alpha}\, \phi(x-\xi)\, \mathrm{d}\xi$$

$$= \frac{\alpha}{\Gamma(1-\alpha)} \int_{0}^{\infty} \left[\phi'(x-\xi) \int_{\xi}^{\infty} \frac{\mathrm{d}\eta}{\eta^{1+\alpha}} \right] \mathrm{d}\xi,$$

so that, interchanging the order of integration

$$D_+^\alpha \, \phi(x) = \frac{\alpha}{\Gamma(1-\alpha)} \int_0^\infty \frac{\phi(x) - \phi(x-\xi)}{\xi^{1+\alpha}} \, \mathrm{d}\xi, \quad 0 < \alpha < 1.$$
(3.3.18)

Here ϕ' denotes the first derivative of ϕ with respect to its argument. The coefficient in front to the integral of (3.3.18) can be rewritten, using known formulas for the Gamma function, as

$$\frac{\alpha}{\Gamma(1-\alpha)} = -\frac{1}{\Gamma(-\alpha)} = \Gamma(1+\alpha) \, \frac{\sin \alpha\pi}{\pi}.$$
(3.3.19)

Similarly, we get for the regressive derivative

$$D_-^\alpha = -\frac{\mathrm{d}}{\mathrm{d}x} \, I_-^{1-\alpha}, \quad 0 < \alpha < 1,$$
(3.3.20)

$$D_-^\alpha \, \phi(x) = \frac{\alpha}{\Gamma(1-\alpha)} \int_0^\infty \frac{\phi(x) - \phi(x+\xi)}{\xi^{1+\alpha}} \, \mathrm{d}\xi, \quad 0 < \alpha < 1.$$
(3.3.21)

Similar results can be given for $\alpha \in (m-1, m)$, $m \in \mathbb{N}$.

3.4. Riesz–Feller Fractional Calculus

The purpose of this section is to combine the Liouville–Weyl fractional integrals and derivatives in order to obtain the pseudo-differential operators considered around the 1950s by Riesz [137] and Feller [44]. In particular the Riesz–Feller fractional derivatives will be used later to generalize the standard diffusion equation by replacing the second-order space derivative. So doing we shall generate all the (symmetric and non-symmetric) Lévy stable probability densities according to our parameterization.

3.4.1. The Riesz fractional integrals and derivatives

The Liouville–Weyl fractional integrals can be combined to give rise to the *Riesz fractional integral* (usually called *Riesz potential*) of order

α, defined as

$$I_0^\alpha \, \phi(x) = \frac{I_+^\alpha \phi(x) + I_-^\alpha \phi(x)}{2\cos(\alpha\pi/2)}$$

$$= \frac{1}{2\,\Gamma(\alpha)\cos(\alpha\pi/2)} \int_{-\infty}^{+\infty} |x - \xi|^{\alpha-1} \, \phi(\xi) \, d\xi, \qquad (3.4.1)$$

for any positive α with the exclusion of odd integer numbers for which $\cos(\alpha\pi/2)$ vanishes. The symbol of the Riesz potential turns out to be

$$\widehat{I_0^\alpha} = |\kappa|^{-\alpha}, \quad \kappa \in \mathbb{R}, \quad \alpha > 0, \quad \alpha \neq 1, 3, 5, \ldots. \qquad (3.4.2)$$

In fact, recalling the symbols of the Liouville–Weyl fractional integrals, see Eq. (3.3.16), we obtain

$$\widehat{I_+^\alpha} + \widehat{I_-^\alpha} = \left[\frac{1}{(-i\kappa)^\alpha} + \frac{1}{(+i\kappa)^\alpha} \right] = \frac{(+i)^\alpha + (-i)^\alpha}{|\kappa|^\alpha} = \frac{2\cos(\alpha\pi/2)}{|\kappa|^\alpha}.$$

We note that, at variance with the Liouville fractional integral, the Riesz potential has the semigroup property only in restricted ranges, e.g.,

$$I_0^\alpha I_0^\beta = I_0^{\alpha+\beta} \quad \text{for } 0 < \alpha < 1, \ 0 < \beta < 1, \ \alpha + \beta < 1. \qquad (3.4.3)$$

From the Riesz potential we can define by analytic continuation the *Riesz fractional derivative* D_0^α, including also the singular case $\alpha = 1$, by formally setting $D_0^\alpha := -I_0^{-\alpha}$, namely, in terms of symbols,

$$\widehat{D_0^\alpha} := -|\kappa|^\alpha. \qquad (3.4.4)$$

We note that the minus sign has been put in order to recover for $\alpha = 2$ the standard second derivative. Indeed, noting that

$$-|\kappa|^\alpha = -(\kappa^2)^{\alpha/2}, \qquad (3.4.5)$$

we recognize that the Riesz fractional derivative of order α is the opposite of the $\alpha/2$-power of the positive definite operator $-\frac{d^2}{dx^2}$

$$D_0^\alpha = -\left(-\frac{d^2}{dx^2} \right)^{\alpha/2}. \qquad (3.4.6)$$

We also note that the two Liouville fractional derivatives are related to the α-power of the first-order differential operator $D = \frac{d}{dx}$. We note that it was Bochner [16] who first introduced the fractional powers of the Laplacian to generalize the diffusion equation.

Restricting our attention to the range $0 < \alpha \leq 2$ the explicit expression for the Riesz fractional derivative turns out to be

$$D_0^\alpha \phi(x) = \begin{cases} -\dfrac{D_+^\alpha \phi(x) + D_-^\alpha \phi(x)}{2 \cos(\alpha\pi/2)} & \text{if } \alpha \neq 1, \\[2mm] -D\,H\,\phi(x) & \text{if } \alpha = 1, \end{cases} \qquad (3.4.7)$$

where H denotes the Hilbert transform operator defined by

$$H\,\phi(x) := \frac{1}{\pi} \int_{-\infty}^{+\infty} \frac{\phi(\xi)}{x - \xi}\,d\xi = \frac{1}{\pi} \int_{-\infty}^{+\infty} \frac{\phi(x - \xi)}{\xi}\,d\xi, \qquad (3.4.8)$$

the integral understood in the Cauchy principal value sense. Incidentally, we note that $H^{-1} = -H$. By using the practical rule (3.3.13) we can derive the symbol of H, namely,

$$\widehat{H} = i\,\text{sign}\,\kappa, \quad \kappa \in \mathbb{R}. \qquad (3.4.9)$$

The expressions in (3.4.7) can be easily verified by manipulating with symbols of "good" operators as below:

$$\widehat{D_0^\alpha} = -\widehat{I_0^{-\alpha}} = -|\kappa|^\alpha = \begin{cases} -\dfrac{(-i\kappa)^\alpha + (+i\kappa)^\alpha}{2 \cos(\alpha\pi/2)} = -|\kappa|^\alpha & \text{if } \alpha \neq 1, \\[2mm] +i\kappa \cdot i\,\text{sign}\,\kappa = -\kappa\,\text{sign}\,\kappa = -|\kappa| & \text{if } \alpha = 1. \end{cases}$$

In particular, from (3.4.7) we recognize that

$$D_0^2 = \frac{1}{2}\left(D_+^2 + D_-^2\right) = \frac{1}{2}\left(\frac{d^2}{dx^2} + \frac{d^2}{dx^2}\right) = \frac{d^2}{dx^2}, \quad \text{but } D_0^1 \neq \frac{d}{dx}.$$

We also recognize that the symbol of D_0^α $(0 < \alpha \leq 2)$ is just the *cumulative function* (logarithm of the characteristic function) of a *symmetric* Lévy stable *pdf*, see, e.g., [45, 145].

We would like to mention the "illuminating" notation introduced by Zaslavsky, see, e.g., [142] to denote our Liouville and Riesz fractional derivatives

$$D_\pm^\alpha = \frac{d^\alpha}{d(\pm x)^\alpha}, \quad D_0^\alpha = \frac{d^\alpha}{d|x|^\alpha}, \quad 0 < \alpha \le 2. \tag{3.4.10}$$

Recalling from (3.4.7) the fractional derivative in Riesz's sense

$$D_0^\alpha \, \phi(x) := -\frac{D_+^\alpha \, \phi(x) + D_-^\alpha \, \phi(x)}{2\cos(\alpha\pi/2)}, \quad 0 < \alpha < 1, \ 1 < \alpha < 2,$$

and using (3.3.18) and (3.3.21) we get for it the following regularized representation, valid also in $\alpha = 1$,

$$D_0^\alpha \, \phi(x) = \Gamma(1+\alpha) \, \frac{\sin(\alpha\pi/2)}{\pi} \int_0^\infty \frac{\phi(x+\xi) - 2\phi(x) + \phi(x-\xi)}{\xi^{1+\alpha}} \, d\xi,$$

$$0 < \alpha < 2. \tag{3.4.11}$$

We note that Eq. (3.4.11) has recently been derived by Gorenflo and Mainardi [58] and improves the corresponding formula in the book by Samko, Kilbas and Marichev [143] which is not valid for $\alpha = 1$.

3.4.2. The Feller fractional integrals and derivatives

A generalization of the Riesz fractional integral and derivative has been proposed by Feller [44] in a pioneering paper, recalled by Samko, Kilbas and Marichev [143], but only recently revised and used by Gorenflo and Mainardi [56]. Feller's intention was indeed to generalize the second-order space derivative entering the standard diffusion equation by a pseudo-differential operator whose symbol is the *cumulative function* (logarithm of the characteristic function) of a general Lévy stable *pdf* according to his parameterization.

Let us now show how to obtain the Feller derivative by inversion of a properly generalized Riesz potential, later called *Feller potential* by Samko, Kilbas and Marichev [143]. Using our notation we define the Feller potential I_θ^α by its symbol obtained from the Riesz potential by a suitable "rotation" by an angle $\theta\,\pi/2$ properly restricted, i.e.,

$$\widehat{I_\theta^\alpha}(\kappa) = |\kappa|^{-\alpha}\, e^{-i\,(\text{sign }\kappa)\,\theta\pi/2}, \quad |\theta| \le \begin{cases} \alpha & \text{if } 0 < \alpha < 1, \\ 2 - \alpha & \text{if } 1 < \alpha \le 2, \end{cases}$$

$$(3.4.12)$$

with $\kappa, \theta \in \mathbb{R}$. Like for the Riesz potential the case $\alpha = 1$ is here omitted. The integral representation of I_θ^α turns out to be

$$I_\theta^\alpha\,\phi(x) = c_-(\alpha,\theta)\, I_+^\alpha\,\phi(x) + c_+(\alpha,\theta)\, I_-^\alpha\,\phi(x), \qquad (3.4.13)$$

where, if $0 < \alpha < 2$, $\alpha \neq 1$,

$$c_+(\alpha,\theta) = \frac{\sin\left[(\alpha - \theta)\,\pi/2\right]}{\sin(\alpha\pi)}, \quad c_-(\alpha,\theta) = \frac{\sin\left[(\alpha + \theta)\,\pi/2\right]}{\sin(\alpha\pi)},$$

$$(3.4.14)$$

and, by passing to the limit (with $\theta = 0$)

$$c_+(2,0) = c_-(2,0) = -1/2. \qquad (3.4.15)$$

In the particular case $\theta = 0$ we get

$$c_+(\alpha,0) = c_-(\alpha,0) = \frac{1}{2\cos(\alpha\pi/2)}, \qquad (3.4.16)$$

and thus, from (3.4.13) and (3.4.16) we recover the Riesz potential (3.4.1). Like the Riesz potential also the Feller potential has the (range-restricted) semi-group property, e.g.,

$$I_\theta^\alpha I_\theta^\beta = I_\theta^{\alpha+\beta} \quad \text{for } 0 < \alpha < 1,\ 0 < \beta < 1,\ \alpha + \beta < 1. \quad (3.4.17)$$

From the Feller potential we can define by analytical continuation the *Feller fractional derivative* D_θ^α, including also the singular case $\alpha = 1$, by setting

$$D_\theta^\alpha := -I_\theta^{-\alpha},$$

so

$$\widehat{D_\theta^\alpha}(\kappa) = -|\kappa|^\alpha\, e^{+i\,(\text{sign }\kappa)\,\theta\pi/2}, \quad |\theta| \le \begin{cases} \alpha & \text{if } 0 < \alpha \le 1, \\ 2 - \alpha & \text{if } 1 < \alpha \le 2. \end{cases}$$

$$(3.4.18)$$

Since for D_θ^α the case $\alpha = 1$ is included, the condition for θ in (3.4.18) can be shortened into

$$|\theta| \leq \min\{\alpha, 2 - \alpha\}, \quad 0 < \alpha \leq 2.$$

We note that the allowed region for the parameters α and θ turns out to be a diamond in the plane $\{\alpha, \theta\}$ with vertices in the points $(0,0)$, $(1,1)$, $(2,0)$, $(1,-1)$.

The representation of $D_\theta^\alpha \phi(x)$ can be obtained from the previous considerations. We have

$$D_\theta^\alpha \phi(x) = \begin{cases} - \left[c_+(\alpha, \theta) D_+^\alpha + c_-(\alpha, \theta) D_-^\alpha \right] \phi(x) & \text{if } \alpha \neq 1, \\ \left[\cos(\theta\pi/2) D_0^1 + \sin(\theta\pi/2) D \right] \phi(x) & \text{if } \alpha = 1. \end{cases}$$
$$(3.4.19)$$

For $\alpha \neq 1$ it is sufficient to note that $c_\mp(-\alpha, \theta) = c_\pm(\alpha, \theta)$. For $\alpha = 1$ we need to recall the symbols of the operators D and $D_0^1 = -DH$, namely $\hat{D} = (-i\kappa)$ and $\widehat{D_0^1} = -|\kappa|$, and note that

$$\widehat{D_\theta^1} = -|\kappa| e^{+i \, (\text{sign } \kappa) \, \theta\pi/2} = - |\kappa| \cos(\theta\pi/2) - (i\kappa) \sin(\theta\pi/2)$$

$$= \cos(\theta\pi/2) \, \widehat{D_0^1} + \sin(\theta\pi/2) \, \hat{D}.$$

We note that in the extremal cases of $\alpha = 1$ we get

$$D_{\pm 1}^1 = \pm D = \pm \frac{d}{dx}. \tag{3.4.20}$$

We also note that the representation by hyper-singular integrals for $0 < \alpha < 2$ (now excluding the cases $\{\alpha = 1, \, \theta \neq 0\}$) can be obtained by using (3.3.18) and (3.3.21) in the first equation of (3.4.19). We get

$$D_\theta^\alpha \phi(x) = \frac{\Gamma(1+\alpha)}{\pi} \left\{ \sin\left[(\alpha + \theta)\pi/2 \right] \int_0^\infty \frac{\phi(x + \xi) - \phi(x)}{\xi^{1+\alpha}} \, d\xi \right.$$

$$\left. + \sin\left[(\alpha - \theta)\pi/2 \right] \int_0^\infty \frac{\phi(x - \xi) - \phi(x)}{\xi^{1+\alpha}} \, d\xi \right\}, \quad (3.4.21)$$

which reduces to (3.4.11) for $\theta = 0$.

For later use we find it convenient to return to the "weight" coefficients $c_\pm(\alpha, \theta)$ in order to outline some properties along with some particular expressions, which can be easily obtained from (3.4.14) with the restrictions on θ given in (3.4.12). We obtain

$$c_\pm \begin{cases} \geq 0 & \text{if } 0 < \alpha < 1, \\ \leq 0 & \text{if } 1 < \alpha \leq 2, \end{cases} \tag{3.4.22}$$

and

$$c_+ + c_- = \frac{\cos(\theta\pi/2)}{\cos(\alpha\pi/2)} \begin{cases} > 0 & \text{if } 0 < \alpha < 1, \\ < 0 & \text{if } 1 < \alpha \leq 2. \end{cases} \tag{3.4.23}$$

In the *extremal* cases we find

$$0 < \alpha < 1, \quad \begin{cases} c_+ = 1, \ c_- = 0 & \text{if } \theta = -\alpha, \\ c_+ = 0, \ c_- = 1 & \text{if } \theta = +\alpha, \end{cases} \tag{3.4.24}$$

$$1 < \alpha < 2, \quad \begin{cases} c_+ = 0, \ c_- = -1 & \text{if } \theta = -(2-\alpha), \\ c_+ = -1, \ c_- = 0 & \text{if } \theta = +(2-\alpha). \end{cases} \tag{3.4.25}$$

In view of the relation of the Feller operators in the framework of *stable* probability density functions, we agree to refer to θ as to the *skewness* parameter.

We must note that in his original paper Feller [44] used a skewness parameter δ different from our θ; the potential introduced by Feller is such that

$$\widehat{I_\delta^\alpha} = \left(|\kappa| e^{-i(\text{sign } \kappa)\delta}\right)^{-\alpha}, \quad \delta = -\frac{\pi}{2}\frac{\theta}{\alpha}, \quad \theta = -\frac{2}{\pi}\alpha\delta. \tag{3.4.26}$$

In their recent book, Uchaikin and Zolotarev [154] have adopted Feller's convention, but using the letter θ for Feller's δ.

3.5. Riemann–Liouville and Caputo Fractional Calculus

In this section, we consider sufficiently well-behaved functions $\psi(t)$ $(t \in \mathbb{R}_0^+)$ with Laplace transform defined as

$$\widetilde{\psi}(s) = \mathcal{L}\{\psi(t); s\} = \int_0^\infty e^{-st}\,\psi(t)\,dt, \quad \Re(s) > a_\psi, \tag{3.5.1}$$

where a_ψ denotes the abscissa of convergence. The inverse Laplace transform is then given as

$$\psi(t) = \mathcal{L}^{-1}\{\widetilde{\psi}(s); t\} = \frac{1}{2\pi i} \int_{Br} e^{st}\, \widetilde{\psi}(s)\, ds, \quad t > 0, \qquad (3.5.2)$$

where Br is a Bromwich path, namely $\{\gamma - i\infty, \gamma + i\infty\}$ with $\gamma > a_\psi$.

It may be convenient to consider $\psi(t)$ as "causal" function in $\mathbb{R}$ namely vanishing for all $t < 0$. For the sake of convenience we adopt the notation $\overset{\mathcal{L}}{\leftrightarrow}$ to denote the juxtaposition of a function with its Laplace transform,[d] with its symbol, namely

$$\psi(t) \overset{\mathcal{L}}{\leftrightarrow} \widetilde{\psi}(s). \qquad (3.5.3)$$

3.5.1. The Riemann–Liouville fractional integrals and derivatives

We first define the Riemann–Liouville (RL) fractional integral and derivative of any order $\alpha > 0$ for a generic (well-behaved) function $\psi(t)$ with $t \in \mathbb{R}^+$.

For the RL *fractional integral* (of order α) we have

$$J^\alpha\, \psi(t) := \frac{1}{\Gamma(\alpha)} \int_0^t (t - \tau)^{\alpha - 1}\, \psi(\tau)\, d\tau, \quad t > 0, \; \alpha > 0. \qquad (3.5.4)$$

For complementation we put $J^0 := \mathbb{I}$ (Identity operator), as it can be justified by passing to the limit $\alpha \to 0$.

The RL integrals possess the semigroup property

$$J^\alpha J^\beta = J^{\alpha + \beta}, \quad \text{for all } \alpha,\, \beta \geq 0. \qquad (3.5.5)$$

The RL *fractional derivative* (of order $\alpha > 0$) is defined as the left-inverse operator of the corresponding RL fractional integral

---

[d]In the ordinary theory of the Laplace transform the condition $\psi(t) \in L_{\text{loc}}(\mathbb{R}^+)$, is necessarily required, and the Bromwich integral is intended in the "principal value" sense. However, we allow an extended use of the theory of Laplace transform which includes Dirac-type generalized functions: then the above integrals must be properly interpreted in the framework of the theory of distributions.

(of order $\alpha > 0$), i.e.,

$$D^\alpha J^\alpha = \mathbb{I}. \tag{3.5.6}$$

Therefore, introducing the positive integer m such that $m - 1 < \alpha \leq m$ and noting that $(D^m J^{m-\alpha}) J^\alpha = D^m (J^{m-\alpha} J^\alpha) = D^m J^m = \mathbb{I}$, we define

$$D^\alpha := D^m J^{m-\alpha}, \quad m - 1 < \alpha \leq m, \tag{3.5.7}$$

namely

$$D^\alpha \psi(t) = \begin{cases} \dfrac{1}{\Gamma(m - \alpha)} \dfrac{\mathrm{d}^m}{\mathrm{d}t^m} \displaystyle\int_0^t \dfrac{\psi(\tau)}{(t - \tau)^{\alpha+1-m}} \, \mathrm{d}\tau, & m - 1 < \alpha < m, \\[2ex] \dfrac{\mathrm{d}^m}{\mathrm{d}t^m} \psi(t), & \alpha = m. \end{cases} \tag{3.5.8}$$

For complementation we put $D^0 := \mathbb{I}$. For $\alpha \to m^-$ we thus recover the standard derivative of order m but the integral formula loses its meaning for $\alpha = m$.

By using the properties of the Eulerian beta and gamma functions it is easy to show the effect of our operators J^α and D^α on the power functions: we have

$$\begin{cases} J^\alpha t^\gamma = \dfrac{\Gamma(\gamma + 1)}{\Gamma(\gamma + 1 + \alpha)} t^{\gamma+\alpha}, \\[2ex] D^\alpha t^\gamma = \dfrac{\Gamma(\gamma + 1)}{\Gamma(\gamma + 1 - \alpha)} t^{\gamma-\alpha}, \end{cases} \quad t > 0, \ \alpha \geq 0, \ \gamma > -1. \tag{3.5.9}$$

These properties are of course a natural generalization of those known when the order is a positive integer.

Note the remarkable fact that the fractional derivative $D^\alpha \psi(t)$ is not zero for the constant function $\psi(t) \equiv 1$ if $\alpha \notin \mathbb{N}$. In fact, the second formula in (3.5.9) with $\gamma = 0$ teaches us that

$$D^\alpha 1 = \frac{t^{-\alpha}}{\Gamma(1 - \alpha)}, \quad \alpha \geq 0, \ t > 0. \tag{3.5.10}$$

This, of course, is $\equiv 0$ for $\alpha \in \mathbb{N}$, due to the poles of the gamma function in the points $0, -1, -2, \ldots$.

3.5.2. The Caputo fractional derivative

We now observe that an alternative definition of the fractional derivative has been introduced in the late sixties by Caputo [21, 22] and soon later adopted in physics to deal long-memory visco-elastic processes by Caputo and Mainardi [30, 31] and, for a more recent review, Mainardi [94].

In this case the fractional derivative, denoted by D_*^α, is defined by exchanging the operators $J^{m-\alpha}$ and D^m in the classical definition (3.5.7), namely

$$D_*^\alpha := J^{m-\alpha} \, D^m, \quad m - 1 < \alpha \le m. \qquad (3.5.11)$$

In the literature, after the appearance in 1999 of the book by Podlubny [133], such derivative is known simply as *Caputo derivative*. Based on (3.5.11) we have

$$D_*^\alpha \, \psi(t) := \begin{cases} \dfrac{1}{\Gamma(m-\alpha)} \displaystyle\int_0^t \dfrac{\psi^{(m)}(\tau)}{(t-\tau)^{\alpha+1-m}} \, \mathrm{d}\tau, & m - 1 < \alpha < m, \\[2mm] \dfrac{\mathrm{d}^m}{\mathrm{d}t^m} \psi(t), & \alpha = m. \end{cases}$$

$$(3.5.12)$$

For $m - 1 < \alpha < m$ the definition (3.5.11) is of course more restrictive than (3.5.7), in that it requires the absolute integrability of the derivative of order m. Whenever we use the operator D_*^α we (tacitly) assume that this condition is met. We easily recognize that in general

$$D^\alpha \, \psi(t) = D^m \, J^{m-\alpha} \, \psi(t) \neq J^{m-\alpha} \, D^m \, \psi(t) = D_*^\alpha \, \psi(t), \quad (3.5.13)$$

unless the function $\psi(t)$ along with its first $m - 1$ derivatives vanishes at $t = 0^+$. In fact, assuming that the passage of the mth derivative under the integral is legitimate, one recognizes that, for

$m - 1 < \alpha < m$ and $t > 0$,

$$D^\alpha \, \psi(t) = D_*^\alpha \, \psi(t) + \sum_{k=0}^{m-1} \frac{t^{k-\alpha}}{\Gamma(k - \alpha + 1)} \, \psi^{(k)}(0^+). \qquad (3.5.14)$$

As noted by Samko, Kilbas and Marichev [143] and Butzer and West-phal [19] the identity (3.5.14) was considered by Liouville himself (but not used for an alternative definition of fractional derivative).

Recalling the fractional derivative of the power functions, see the second equation in (3.5.9), we can rewrite (3.5.14) in the equivalent form

$$D^\alpha \left(\psi(t) - \sum_{k=0}^{m-1} \frac{t^k}{k!} \, \psi^{(k)}(0^+) \right) = D_*^\alpha \, \psi(t). \qquad (3.5.15)$$

The subtraction of the Taylor polynomial of degree $m - 1$ at $t = 0^+$ from $\psi(t)$ means a sort of regularization of the fractional derivative. In particular, according to this definition, the relevant property for which the fractional derivative of a constant is still zero can be easily recognized,

$$D_*^\alpha 1 \equiv 0, \quad \alpha > 0. \qquad (3.5.16)$$

As a consequence of (3.5.15) we can interpret the Caputo derivative as a sort of regularization of the RL derivative as soon as the values $\psi^k(0^+)$ are finite; in this sense such fractional derivative was inde-pendently introduced in 1968 by Dzherbashyan and Nersesian [40], as pointed out in interesting papers by Kochubei [78, 79]. In this respect the regularized fractional derivative is sometimes referred to as the *Caputo–Dzherbashyan derivative*.

We now explore the most relevant differences between the two fractional derivatives (3.5.7) and (3.5.11). We agree to denote (3.5.11) as the *Caputo fractional derivative* to distinguish it from the standard A–R fractional derivative (3.5.7). We observe, again by looking at second equation in (3.5.9), that $D^\alpha t^{\alpha-k} \equiv 0$, $t > 0$ for $\alpha > 0$, and $k = 1, 2, \ldots, m$. We thus recognize the following statements about functions which for $t > 0$ admit the same fractional derivative of order α, with $m - 1 < \alpha \le m$, $m \in \mathbb{N}$,

$$D^\alpha \psi(t) = D^\alpha \phi(t) \Leftrightarrow \psi(t) = \phi(t) + \sum_{j=1}^{m} c_j t^{\alpha-j}, \qquad (3.5.17)$$

$$D_*^\alpha \psi(t) = D_*^\alpha \phi(t) \Leftrightarrow \psi(t) = \phi(t) + \sum_{j=1}^{m} c_j t^{m-j}, \qquad (3.5.18)$$

where the coefficients c_j are arbitrary constants.

For the two definitions we also note a difference with respect to the *formal* limit as $\alpha \to (m-1)^+$; from (3.5.7) and (3.5.11) we obtain respectively,

$$D^\alpha \psi(t) \to D^m J \psi(t) = D^{m-1} \psi(t), \qquad (3.5.19)$$

$$D_*^\alpha \psi(t) \to J D^m \psi(t) = D^{m-1} \psi(t) - \psi^{(m-1)}(0^+). \qquad (3.5.20)$$

We now consider the *Laplace transform* of the two fractional derivatives. For the A–R fractional derivative D^α, the Laplace transform, assumed to exist, requires the knowledge of the (bounded) initial values of the fractional integral $J^{m-\alpha}$ and of its integer derivatives of order $k = 1, 2, \ldots, m - 1$. The corresponding rule reads, in our notation,

$$D^\alpha \psi(t) \overset{\mathcal{L}}{\leftrightarrow} s^\alpha \widetilde{\psi}(s) - \sum_{k=0}^{m-1} D^k J^{(m-\alpha)} \psi(0^+) s^{m-1-k}, \quad m-1 < \alpha \le m.$$
$$(3.5.21)$$

For the *Caputo* fractional derivative the Laplace transform technique requires the knowledge of the (bounded) initial values of the function and of its integer derivatives of order $k = 1, 2, \ldots, m - 1$, in analogy with the case when $\alpha = m$. In fact, noting that $J^\alpha D_*^\alpha = J^\alpha J^{m-\alpha} D^m = J^m D^m$, we have

$$J^\alpha D_*^\alpha \psi(t) = \psi(t) - \sum_{k=0}^{m-1} \psi^{(k)}(0^+) \frac{t^k}{k!}, \qquad (3.5.22)$$

so we easily prove the following rule for the Laplace transform,

$$D_*^\alpha \psi(t) \overset{\mathcal{L}}{\leftrightarrow} s^\alpha \widetilde{\psi}(s) - \sum_{k=0}^{m-1} \psi^{(k)}(0^+)\, s^{\alpha-1-k}, \quad m-1 < \alpha \le m.$$

$$(3.5.23)$$

Indeed the result (3.5.23), first stated by Caputo [22], appears as the "natural" generalization of the corresponding well-known result for $\alpha = m$.

Gorenflo and Mainardi (1997) [55] have pointed out the major utility of the Caputo fractional derivative in the treatment of differential equations of fractional order for *physical applications*. In fact, in physical problems, the initial conditions are usually expressed in terms of a given number of bounded values assumed by the field variable and its derivatives of integer order, despite the fact that the governing evolution equation may be a generic integro-differential equation and therefore, in particular, a fractional differential equation.

We note that the Caputo fractional derivative is not mentioned in the standard books of fractional calculus (including the encyclopedic treatise by Samko, Kilbas and Marichev [143]) with the exception of the recent book by Podlubny [133], where this derivative is largely treated in the theory and applications. Several applications have also been treated by Caputo himself from the seventies up to nowadays, see, e.g., [21–29].

3.6. Mittag-Leffler Functions

The Mittag-Leffler function is so named after the great Swedish mathematician Gosta Mittag-Leffler, who introduced and investigated it at the beginning of the 20th century in a sequence of five notes [124–128]. In this section, we shall consider the Mittag-Leffler function and some of the related functions which are relevant for fractional evolution processes. It is our intention to provide a short reference-historical background and a review of the main properties of these functions, based on our papers, see [89–111].

3.6.1. Reference-historical background

We note that the Mittag-Leffler type functions, being ignored in the common books on special functions, are unknown to the majority of scientists. Even in the 1991 Mathematics Subject Classification these functions cannot be found. However they have now appeared in the new MSC scheme of the year 2000 under the number 33E12 ("Mittag-Leffler functions").[e]

A description of the most important properties of these functions (with relevant references up to the fifties) can be found in the third volume of the Bateman Project [42], in the chapter *XVIII* on *Miscellaneous Functions*. The treatises where great attention is devoted to them are those by Djrbashian (or Dzherbashian) [37, 38] We also recommend the classical treatise on complex functions by Sansone and Gerretsen [144]. The Mittag-Leffler functions are widely used in the books on fractional calculus and its applications, see, e.g., [76, 88, 133, 143, 156].

Since the times of Mittag-Leffler several scientists have recognized the importance of the functions named after him, providing interesting mathematical and physical applications which unfortunately are not much known. As pioneering works of mathematical nature in the field of fractional integral and differential equations, we like to quote those by Hille and Tamarkin [67] and Barret [13]. Hille and Tamarkin have provided the solution of the Abel integral equation of the second kind in terms of a Mittag-Leffler function, whereas Barret has expressed the general solution of the linear fractional differential equation with constant coefficients in terms of Mittag-Leffler functions. As former applications in physics we like to quote the contributions by Cole [33] in connection with nerve conduction, see also Davis [35], and by Gross [60] in connection with mechanical

---

[e]More details on the Mittag-Leffler functions can be found in the 2011 survey by Haubold, Mathai and Saxena [63] and in the more recent monograph, here added in proof, R. Gorenflo, A.A Kilbas, F. Mainardi and S.V. Rogosin, *Mittag-Leffler Functions. Related Topics and Applications*, Springer, Berlin, 2014.

relaxation. Subsequently, Caputo and Mainardi [30, 31] have shown that Mittag-Leffler functions are present whenever derivatives of fractional order are introduced in the constitutive equations of a linear viscoelastic body. Since then, several other authors have pointed out the relevance of these functions for fractional viscoelastic models, see, e.g., [94].

3.6.2. The Mittag-Leffler functions $E_\alpha(z), E_{\alpha,\beta}(z)$

The Mittag-Leffler function $E_\alpha(z)$ with $\alpha > 0$ is defined by its power series, which converges in the whole complex plane,

$$E_\alpha(z) := \sum_{n=0}^{\infty} \frac{z^n}{\Gamma(\alpha n + 1)}, \quad \alpha > 0, \ z \in \mathbb{C}. \tag{3.6.1}$$

It turns out that $E_\alpha(z)$ is an *entire function* of order $\rho = 1/\alpha$ and type 1. This property is still valid but with $\rho = 1/\Re(\alpha)$, if $\alpha \in \mathbb{C}$ with *positive real part*, as formerly noted by Mittag-Leffler himself in [126].

The Mittag-Leffler function provides a simple generalization of the exponential function to which it reduces for $\alpha = 1$. Other particular cases of (1), from which elementary functions are recovered, are

$$E_2(+z^2) = \cosh z, \quad E_2(-z^2) = \cos z, \ z \in \mathbb{C}, \tag{3.6.2}$$

and

$$E_{1/2}(\pm z^{1/2}) = e^z \left[1 + \mathrm{erf}\,(\pm z^{1/2})\right] = e^z \,\mathrm{erfc}\,(\mp z^{1/2}), \quad z \in \mathbb{C}, \tag{3.6.3}$$

where erf (erfc) denotes the (complementary) error function defined as

$$\mathrm{erf}\,(z) := \frac{2}{\sqrt{\pi}} \int_0^z e^{-u^2} \, du, \quad \mathrm{erfc}\,(z) := 1 - \mathrm{erf}\,(z), \quad z \in \mathbb{C}. \tag{3.6.4}$$

In (3.6.4) by $z^{1/2}$ we mean the principal value of the square root of z in the complex plane cut along the negative real semi-axis. With this choice $\pm z^{1/2}$ turns out to be positive/negative for $z \in \mathbb{R}^+$. A straightforward generalization of the Mittag-Leffler function, originally due to Agarwal in 1953 based on a note by Humbert, see [69–71], is obtained by replacing the additive constant 1 in the argument of the Gamma function in (3.6.1) by an arbitrary complex parameter β. For the generalized Mittag-Leffler function we agree to use the notation

$$E_{\alpha,\beta}(z) := \sum_{n=0}^{\infty} \frac{z^n}{\Gamma(\alpha n + \beta)}, \quad \alpha > 0, \ \beta \in \mathbb{C}, \ z \in \mathbb{C}. \quad (3.6.5)$$

Particular simple cases are

$$E_{1,2}(z) = \frac{e^z - 1}{z}, \quad E_{2,2}(z) = \frac{\sinh(z^{1/2})}{z^{1/2}}. \quad (3.6.6)$$

We note that $E_{\alpha,\beta}(z)$ is still an entire function of order $\rho = 1/\alpha$ and type 1.

3.6.3. The Mittag-Leffler integral representation and asymptotic expansions

Many of the important properties of $E_\alpha(z)$ follow from Mittag-Leffler's *integral representation*

$$E_\alpha(z) = \frac{1}{2\pi i} \int_{Ha} \frac{\zeta^{\alpha-1} e^\zeta}{\zeta^\alpha - z} \, d\zeta, \quad \alpha > 0, \ z \in \mathbb{C}, \quad (3.6.7)$$

where the path of integration Ha (the *Hankel path*) is a loop which starts and ends at $-\infty$ and encircles the circular disk $|\zeta| \leq |z|^{1/\alpha}$ in the positive sense: $-\pi \leq \arg\zeta \leq \pi$ on Ha. To prove (3.6.7), expand the integrand in powers of ζ, integrate term-by-term, and use Hankel's integral for the reciprocal of the Gamma function.

The integrand in (3.6.7) has a branch-point at $\zeta = 0$. The complex ζ-plane is cut along the negative real semi-axis, and in the cut plane the integrand is single-valued: the principal branch of ζ^α

is taken in the cut plane. The integrand has poles at the points $\zeta_m = z^{1/\alpha}\, e^{2\pi i m/\alpha}$, m integer, but only those of the poles lie in the cut plane for which $-\alpha\pi < \arg z + 2\pi m < \alpha\pi$. Thus, the number of the poles inside Ha is either $[\alpha]$ or $[\alpha+1]$, according to the value of $\arg z$.

The most interesting properties of the Mittag-Leffler function are associated with its asymptotic developments as $z \to \infty$ in various sectors of the complex plane. These properties can be summarized as follows. For the case $0 < \alpha < 2$ we have

$$E_\alpha(z) \sim \frac{1}{\alpha}\exp(z^{1/\alpha}) - \sum_{k=1}^{\infty}\frac{z^{-k}}{\Gamma(1-\alpha k)}, \qquad |z| \to \infty, \ |\arg z| < \alpha\pi/2,$$

(3.6.8a)

$$E_\alpha(z) \sim -\sum_{k=1}^{\infty}\frac{z^{-k}}{\Gamma(1-\alpha k)}, \qquad |z| \to \infty, \ \alpha\pi/2 < \arg z < 2\pi - \alpha\pi/2.$$

(3.6.8b)

For the case $\alpha \geq 2$ we have

$$E_\alpha(z) \sim \frac{1}{\alpha}\sum_{m}\exp(z^{1/\alpha}e^{2\pi i m/\alpha}) - \sum_{k=1}^{\infty}\frac{z^{-k}}{\Gamma(1-\alpha k)}, \qquad |z| \to \infty,$$

(3.6.9)

where m takes all integer values such that $-\alpha\pi/2 < \arg z + 2\pi m < \alpha\pi/2$, and $\arg z$ can assume any value from $-\pi$ to $+\pi$.

From the asymptotic properties (3.6.8)–(3.6.9) and the definition of the order of an entire function, we infer that the Mittag-Leffler function is an *entire function of order* $1/\alpha$ for $\alpha > 0$; in a certain sense each $E_\alpha(z)$ is the simplest entire function of its order. The Mittag-Leffler function also furnishes examples and counterexamples for the growth and other properties of entire functions of finite order.

Finally, the integral representation for the generalized Mittag-Leffler function reads

$$E_{\alpha,\beta}(z) = \frac{1}{2\pi i}\int_{Ha}\frac{\zeta^{\alpha-\beta}\,e^{\zeta}}{\zeta^{\alpha}-z}\,d\zeta, \qquad \alpha > 0, \ \beta \in \mathbb{C}, \ z \in \mathbb{C}. \quad (3.6.10)$$

3.6.4. The Laplace transform pairs related to the Mittag-Leffler functions

The Mittag-Leffler functions are connected to the Laplace integral through the equation

$$\int_0^\infty e^{-u} E_\alpha \left(u^\alpha z\right) \, \mathrm{d}u = \frac{1}{1-z}, \quad \alpha > 0. \tag{3.6.11}$$

The integral at the left-hand side was evaluated by Mittag-Leffler who showed that the region of its convergence contains the unit circle and is bounded by the line $\operatorname{Re} z^{1/\alpha} = 1$. Putting in (3.6.11) $u = st$ and $u^\alpha z = -a\,t^\alpha$ with $t \geq 0$ and $a \in \mathbb{C}$, and using the sign $\overset{\mathcal{L}}{\leftrightarrow}$ for the juxtaposition of a function depending on t with its Laplace transform depending on s, we get the following Laplace transform pairs

$$E_\alpha \left(-a\,t^\alpha\right) \overset{\mathcal{L}}{\leftrightarrow} \frac{s^{\alpha-1}}{s^\alpha + a}, \quad \Re(s) > |a|^{1/\alpha}. \tag{3.6.12}$$

More generally one can show

$$\int_0^\infty e^{-u}\, u^{\beta-1}\, E_{\alpha,\beta}\left(u^\alpha z\right) \, \mathrm{d}u = \frac{1}{1-z}, \quad \alpha, \beta > 0, \tag{3.6.13}$$

and

$$t^{\beta-1}\, E_{\alpha,\beta}\left(a\,t^\alpha\right) \overset{\mathcal{L}}{\leftrightarrow} \frac{s^{\alpha-\beta}}{s^\alpha - a}, \quad \Re(s) > |a|^{1/\alpha}. \tag{3.6.14}$$

We note that the results (3.6.12) and (3.6.14) were used by Humbert [70] to obtain a number of functional relations satisfied by $E_\alpha(z)$ and $E_{\alpha,\beta}(z)$.

3.6.5. Fractional relaxation and fractional oscillation

In our CISM Lecture notes, see [55], we have worked out the key role of the Mittag-Leffler type functions $E_\alpha\left(-a\,t^\alpha\right)$ in treating Abel integral equations of the second kind and fractional differential equations, so improving the former results by Hille and Tamarkin [67] and Barret [13], respectively. In particular, assuming $a > 0$, we have discussed the peculiar characters of these functions (*power-law decay*) for $0 < \alpha < 1$ and for $1 < \alpha < 2$ related to *fractional relaxation* and *fractional oscillation* processes, respectively, see also [93] and [54].

Generally speaking, we consider the following differential equation of fractional order $\alpha > 0$,

$$D_*^\alpha u(t) = D^\alpha \left(u(t) - \sum_{k=0}^{m-1} \frac{t^k}{k!} u^{(k)}(0^+) \right) = -u(t) + q(t), \quad t > 0,$$
(3.6.15)

where $u = u(t)$ is the field variable and $q(t)$ is a given function, continuous for $t \geq 0$. Here m is a positive integer uniquely defined by $m - 1 < \alpha \leq m$, which provides the number of the prescribed initial values $u^{(k)}(0^+) = c_k$, $k = 0, 1, 2, \ldots, m - 1$. Implicit in the form of (3.6.15) is our desire to obtain solutions $u(t)$ for which the $u^{(k)}(t)$ are continuous for $t \geq 0$ for $k = 0, 1, 2, \ldots, m - 1$. In particular, the cases of *fractional relaxation* and *fractional oscillation* are obtained for $0 < \alpha < 1$ and $1 < \alpha < 2$, respectively.

The application of the Laplace transform through the Caputo formula (3.5.19) yields

$$\widetilde{u}(s) = \sum_{k=0}^{m-1} c_k \frac{s^{\alpha-k-1}}{s^\alpha + 1} + \frac{1}{s^\alpha + 1} \widetilde{q}(s).$$
(3.6.16)

Then, using (3.6.12), we put for $k = 0, 1, \ldots, m - 1$,

$$u_k(t) := J^k e_\alpha(t) \overset{\mathcal{L}}{\leftrightarrow} \frac{s^{\alpha-k-1}}{s^\alpha + 1}, \quad e_\alpha(t) := E_\alpha(-t^\alpha) \overset{\mathcal{L}}{\leftrightarrow} \frac{s^{\alpha-1}}{s^\alpha + 1},$$
(3.6.17)

and, from inversion of the Laplace transforms in (3.6.16), using $u_0(0^+) = 1$, we find

$$u(t) = \sum_{k=0}^{m-1} c_k u_k(t) - \int_0^t q(t-\tau) u_0'(\tau) \, d\tau.$$
(3.6.18)

In particular, the formula (3.6.18) encompasses the solutions for $\alpha = 1, 2$, since $e_1(t) = \exp(-t)$, $e_2(t) = \cos t$. When α is not integer, namely for $m - 1 < \alpha < m$, we note that $m - 1$ represents the integer part of α (usually denoted by $[\alpha]$) and m the number of initial conditions necessary and sufficient to ensure the uniqueness of the solution $u(t)$. Thus the m-functions $u_k(t) = J^k e_\alpha(t)$ with $k = 0, 1, \ldots, m-1$ represent those particular solutions of the *homogeneous* equation which satisfy the initial conditions $u_k^{(h)}(0^+) = \delta_{kh}$, $h, k = 0, 1, \ldots, m-1$, and therefore they represent the *fundamental solutions* of the fractional equation (3.6.15), in analogy with the case $\alpha = m$. Furthermore, the function $u_\delta(t) = -u_0'(t) = -e_\alpha'(t)$ represents the *impulse-response solution*.

We have derived the relevant properties of the basic functions $e_\alpha(t)$ directly from their representation as a Laplace inverse integral

$$e_\alpha(t) = \frac{1}{2\pi i} \int_{Br} e^{st} \frac{s^{\alpha-1}}{s^\alpha + 1} \, ds, \qquad (3.6.19)$$

in detail for $0 < \alpha \leq 2$, without detouring on the general theory of Mittag-Leffler functions in the complex plane. In (3.6.19) Br denotes a Bromwich path, i.e., a line $\Re(s) = \sigma$ with a value $\sigma \geq 1$ and $\operatorname{Im} s$ running from $-\infty$ to $+\infty$.

For transparency reasons, we separately discuss the cases (a) $0 < \alpha < 1$ and (b) $1 < \alpha < 2$, recalling that in the limiting cases $\alpha = 1, 2$, we know $e_\alpha(t)$ as elementary function, namely $e_1(t) = e^{-t}$ and $e_2(t) = \cos t$. For α not integer the power function s^α is uniquely defined as $s^\alpha = |s|^\alpha e^{i \arg s}$, with $-\pi < \arg s < \pi$, that is in the complex s-plane cut along the negative real axis.

The essential step consists in decomposing $e_\alpha(t)$ into two parts according to $e_\alpha(t) = f_\alpha(t) + g_\alpha(t)$, as indicated below. In case (a) the function $f_\alpha(t)$, in case (b) the function $-f_\alpha(t)$ is *completely monotone*; in both cases $f_\alpha(t)$ tends to zero as t tends to infinity, from above in case (a), from below in case (b). The other part, $g_\alpha(t)$, is identically vanishing in case (a), but of *oscillatory* character with exponentially decreasing amplitude in case (b).

For the oscillatory part we obtain via the residue theorem of complex analysis

$$g_\alpha(t) = \frac{2}{\alpha} e^{t \cos(\pi/\alpha)} \cos\left[t \sin\left(\frac{\pi}{\alpha}\right)\right] \quad \text{if } 1 < \alpha < 2. \qquad (3.6.20)$$

We note that this function exhibits oscillations with circular frequency $\omega(\alpha) = \sin(\pi/\alpha)$ and with an exponentially decaying amplitude with rate $\lambda(\alpha) = |\cos(\pi/\alpha)| = -\cos(\pi/\alpha)$.

For the monotonic part we obtain

$$f_\alpha(t) := \int_0^\infty e^{-rt} K_\alpha(r) \, dr, \qquad (3.6.21)$$

with

$$K_\alpha(r) = -\frac{1}{\pi} \operatorname{Im} \left\{ \left. \frac{s^{\alpha-1}}{s^\alpha + 1} \right|_{s=r\,e^{i\pi}} \right\} = \frac{1}{\pi} \frac{r^{\alpha-1} \sin(\alpha\pi)}{r^{2\alpha} + 2\,r^\alpha \cos(\alpha\pi) + 1}. \qquad (3.6.22)$$

This function $K_\alpha(r)$ vanishes identically if α is an integer, it is positive for all r if $0 < \alpha < 1$, negative for all r if $1 < \alpha < 2$. In fact in (3.6.22) the denominator is, for α not integer, always positive being greater than $(r^\alpha - 1)^2 \geq 0$. Hence $f_\alpha(t)$ has the aforementioned monotonicity properties, decreasing towards zero in case (a), increasing towards zero in case (b). We also note that, in order to satisfy the initial condition $e_\alpha(0^+) = 1$, we find $\int_0^\infty K_\alpha(r) \, dr = 1$ if $0 < \alpha < 1$, $\int_0^\infty K_\alpha(r) \, dr = 1 - 2/\alpha$ if $1 < \alpha < 2$.

In addition to the basic fundamental solutions, $u_0(t) = e_\alpha(t)$, we need to compute the impulse-response solutions $u_\delta(t) = -D^1 e_\alpha(t)$ for cases (a) and (b) and, only in case (b), the second fundamental solution $u_1(t) = J^1 e_\alpha(t)$.

For this purpose we note that in general it turns out that

$$J^k f_\alpha(t) = \int_0^\infty e^{-rt} K_\alpha^k(r) \, dr, \qquad (3.6.23)$$

with

$$K_\alpha^k(r) := (-1)^k \, r^{-k} K_\alpha(r) = \frac{(-1)^k}{\pi} \frac{r^{\alpha-1-k} \sin(\alpha\pi)}{r^{2\alpha} + 2\,r^\alpha \cos(\alpha\pi) + 1}, \qquad (3.6.24)$$

where $K_\alpha(r) = K_\alpha^0(r)$, and

$$J^k g_\alpha(t) = \frac{2}{\alpha} e^{t \cos(\pi/\alpha)} \cos\left[t \sin\left(\frac{\pi}{\alpha}\right) - k\frac{\pi}{\alpha} \right]. \qquad (3.6.25)$$

This can be done in direct analogy to the computation of the functions $e_\alpha(t)$, the Laplace transform of $J^k e_\alpha(t)$ being given by (3.6.17). For the impulse-response solution we note that the effect of the differential operator D^1 is the same as that of the virtual operator J^{-1}.

In conclusion we can resume the solutions for the fractional relaxation and oscillation equations as follows:

(a) $0 < \alpha < 1$,

$$u(t) = c_0 \, u_0(t) + \int_0^t q(t - \tau) \, u_\delta(\tau) \, d\tau, \qquad (3.6.26a)$$

where

$$\begin{cases} u_0(t) = \displaystyle\int_0^\infty e^{-rt} \, K_\alpha^0(r) \, dr, \\[2mm] u_\delta(t) = -\displaystyle\int_0^\infty e^{-rt} \, K_\alpha^{-1}(r) \, dr, \end{cases} \qquad (3.6.27a)$$

with $u_0(0^+) = 1$, $u_\delta(0^+) = \infty$;

(b) $1 < \alpha < 2$,

$$u(t) = c_0 \, u_0(t) + c_1 \, u_1(t) + \int_0^t q(t - \tau) \, u_\delta(\tau) \, d\tau, \qquad (3.6.26b)$$

where

$$\begin{cases} u_0(t) = \displaystyle\int_0^\infty e^{-rt} \, K_\alpha^0(r) \, dr + \frac{2}{\alpha} e^{t \cos(\pi/\alpha)} \, \cos\left[t \sin\left(\frac{\pi}{\alpha}\right) \right], \\[3mm] u_1(t) = \displaystyle\int_0^\infty e^{-rt} \, K_\alpha^1(r) \, dr \\[2mm] \qquad\quad + \dfrac{2}{\alpha} e^{t \cos(\pi/\alpha)} \, \cos\left[t \sin\left(\dfrac{\pi}{\alpha}\right) - \dfrac{\pi}{\alpha} \right], \\[3mm] u_\delta(t) = -\displaystyle\int_0^\infty e^{-rt} \, K_\alpha^{-1}(r) \, dr \\[2mm] \qquad\quad - \dfrac{2}{\alpha} e^{t \cos(\pi/\alpha)} \, \cos\left[t \sin\left(\dfrac{\pi}{\alpha}\right) + \dfrac{\pi}{\alpha} \right], \end{cases}$$

$$\qquad\qquad\qquad\qquad\qquad\qquad\qquad\qquad\qquad\qquad (3.6.27b)$$

with $u_0(0^+) = 1$, $u_0'(0^+) = 0$, $u_1(0^+) = 0$, $u_1'(0^+) = 1$, $u_\delta(0^+) = 0$, $u_\delta'(0^+) = +\infty$. We have verified that our present results confirm those obtained by Blank [15] by a numerical calculations and those obtained by Mainardi [92] by an analytical treatment, valid when α is a rational number, see later. Of particular interest is the case $\alpha = 1/2$ where we recover a well-known formula of the Laplace transform theory, see e.g., [41]

$$e_{1/2}(t) := E_{1/2}(-\sqrt{t}) = e^t \operatorname{erfc}(\sqrt{t}) \overset{\mathcal{L}}{\leftrightarrow} \frac{1}{s^{1/2}\left(s^{1/2} + 1\right)},$$
$$(3.6.28)$$

where erfc denotes the *complementary error* function defined in (3.6.4). Explicitly we have

$$E_{1/2}(-\sqrt{t}) = e^t \frac{2}{\sqrt{\pi}} \int_{\sqrt{t}}^\infty e^{-u^2}\, du. \qquad (3.6.29)$$

We now want to point out that in both the cases (a) and (b) (in which α is just not integer), i.e., for *fractional relaxation* and *fractional oscillation*, all the fundamental and impulse-response solutions exhibit an *algebraic decay* as $t \to \infty$, as discussed below. This algebraic decay is the most important effect of the non-integer derivative in our equations, which dramatically differs from the *exponential decay* present in the ordinary relaxation and damped-oscillation phenomena.

Let us start with the asymptotic behavior of $u_0(t)$. To this purpose we first derive an asymptotic series for the function $f_\alpha(t)$, valid for $t \to \infty$. We then consider the spectral representation (3.6.21)–(3.6.22) and expand the spectral function for small r. Then the Watson lemma yields

$$f_\alpha(t) = \sum_{n=1}^N (-1)^{n-1} \frac{t^{-n\alpha}}{\Gamma(1 - n\alpha)} + O(t^{-(N+1)\alpha}), \quad \text{as } t \to \infty.$$
$$(3.6.30)$$

We note that this asymptotic expansion coincides with that for $u_0(t) = e_\alpha(t)$, having assumed $0 < \alpha < 2$ ($\alpha \neq 1$). In fact the contribution of $g_\alpha(t)$ is identically zero if $0 < \alpha < 1$ and exponentially small as $t \to \infty$ if $1 < \alpha < 2$.

The asymptotic expansions of the solutions $u_1(t)$ and $u_\delta(t)$ are obtained from (3.6.30) integrating or differentiating term-by-term with respect to t. Taking the leading term of the asymptotic expansions, we obtain the asymptotic representations of the solutions $u_0(t)$, $u_1(t)$ and $u_\delta(t)$ as $t \to \infty$,

$$u_0(t) \sim \frac{t^{-\alpha}}{\Gamma(1-\alpha)}, \quad u_1(t) \sim \frac{t^{1-\alpha}}{\Gamma(2-\alpha)}, \quad u_\delta(t) \sim -\frac{t^{-\alpha-1}}{\Gamma(-\alpha)},$$

(3.6.31)

that point out the algebraic decay.

We would like to remark the difference between fractional relaxation governed by the Mittag-Leffler type function $E_\alpha(-at^\alpha)$ and stretched relaxation governed by a stretched exponential function $\exp(-bt^\alpha)$ with $\alpha, a, b > 0$ for $t \geq 0$. A common behavior is achieved only in a restricted range $0 \leq t \ll 1$ where we can have

$$E_\alpha(-at^\alpha) \simeq 1 - \frac{a}{\Gamma(\alpha+1)} t^\alpha$$

$$= 1 - bt^\alpha \simeq e^{-bt^\alpha} \quad \text{if } b = \frac{a}{\Gamma(\alpha+1)}.$$

(3.6.32)

3.6.6. Other formulas: summation and integration

For completeness hereafter we exhibit some formulas related to summation and integration of ordinary Mittag-Leffler functions (in one parameter α), referring the interested reader to [63, 133] for their proof and for their generalizations to two parameters.

Concerning summation we outline

$$E_\alpha(z) = \frac{1}{p} \sum_{h=0}^{p-1} E_{\alpha/p}(z^{1/p} e^{i2\pi h/p}), \quad p \in \mathbb{N},$$

(3.6.33)

from which we derive the *duplication formula*

$$E_\alpha(z) = \tfrac{1}{2}[E_{\alpha/2}(+z^{1/2}) + E_{\alpha/2}(-z^{1/2})].$$

(3.6.34)

As an example of this formula we can recover, for $\alpha = 2$, the expressions of $\cosh z$ and $\cos z$ in terms of two exponential functions.

Concerning integration we outline another interesting *duplication formula*

$$E_{\alpha/2}(-t^{\alpha/2}) = \frac{1}{\sqrt{\pi\,t}} \int_0^\infty e^{-x^2/(4t)}\, E_\alpha(-x^\alpha)\, dx, \quad x > 0,\ t > 0.$$

$$(3.6.35)$$

It can be derived by applying a theorem of the Laplace transform theory (known as Efros theorem).

3.6.7. The Mittag-Leffler functions of rational order

Let us now consider the Mittag-Leffler functions of rational order $\alpha = p/q$ with $p,\ q \in \mathbb{N}$, relatively prime. The relevant functional relations, that we quote from [37; 42, Vol. 3], turn out to be

$$\left(\frac{d}{dz}\right)^p E_p(z^p) = E_p(z^p),\qquad (3.6.36)$$

$$\frac{d^p}{dz^p} E_{p/q}(z^{p/q}) = E_{p/q}(z^{p/q}) + \sum_{k=1}^{q-1} \frac{z^{-kp/q}}{\Gamma(1 - k\,p/q)}, \quad q = 2, 3, \ldots,$$

$$(3.6.37)$$

$$E_{p/q}(z) = \frac{1}{p} \sum_{h=0}^{p-1} E_{1/q}(z^{1/p} e^{i2\pi h/p}), \qquad (3.6.38)$$

and

$$E_{1/q}(z^{1/q}) = e^z \left[1 + \sum_{k=1}^{q-1} \frac{\gamma(1 - k/q, z)}{\Gamma(1 - k/q)} \right], \quad q = 2, 3, \ldots, \quad (3.6.39)$$

where $\gamma(a, z) := \int_0^z e^{-u}\, u^{a-1}\, du$ denotes the *incomplete gamma function* defined as

$$\gamma(a, z) := \int_0^z e^{-u}\, u^{a-1}\, du. \qquad (3.6.40)$$

One easily recognizes that the relations (3.6.36) and (3.6.37) are immediate consequences of the definition (3.6.1).

The relation (3.6.39) shows how the Mittag-Leffler functions of rational order can be expressed in terms of exponentials and incomplete gamma functions. In particular, taking in (3.6.39) $q = 2$, we now can verify again the relation (3.6.4). In fact, from (3.6.39) we obtain

$$E_{1/2}(z^{1/2}) = e^z \left[1 + \frac{1}{\sqrt{\pi}} \gamma(1/2,\, z) \right], \qquad (3.6.41)$$

which is equivalent to (3.6.4) if we use the relation

$$\operatorname{erf}(z) = \gamma(1/2, z^2)/\sqrt{\pi}, \qquad (3.6.42)$$

see, e.g., [42, Vol. 1; 3].

3.6.8. Some plots of the Mittag-Leffler functions

For readers' convenience we now consider the functions

$$\psi_\alpha(t) := E_\alpha(-t^\alpha), \quad t \geq 0,\ 0 < \alpha < 1, \qquad (3.6.43)$$

and

$$\phi_\alpha(t) := t^{-(1-\alpha)} E_{\alpha,\alpha}(-t^\alpha)$$
$$= -\frac{\mathrm{d}}{\mathrm{d}t} E_\alpha(-t^\alpha), \quad t \geq 0,\ 0 < \alpha < 1, \qquad (3.6.44)$$

that play fundamental roles in fractional relaxation. The plots of $\psi_\alpha(t)$ and $\phi_\alpha(t)$ are shown in Figs. 3.1 and 3.2, respectively, for some rational values of the parameter α, by adopting linear and logarithmic scales. It is evident that for $\alpha \to 1^-$ the two functions reduce to the standard exponential function $\exp(-t)$.

It is worth noting the algebraic decay of $\psi_\alpha(t)$ and $\phi_\alpha(t)$ as $t \to \infty$:

$$\psi_\alpha(t) \sim \frac{\sin(\alpha\pi)}{\pi} \frac{\Gamma(\alpha)}{t^\alpha}, \quad \phi_\alpha(t) \sim \frac{\sin(\alpha\pi)}{\pi} \frac{\Gamma(\alpha+1)}{t^{(\alpha+1)}}, \quad t \to +\infty. \qquad (3.6.45)$$

3.7. Wright Functions

The *Wright* function that we denote by $\Phi_{\lambda,\mu}(z), z \in \mathbb{C}$, with the parameters $\lambda > -1$ and $\mu \in \mathbb{C}$, is so named after the British mathematician E. Maitland Wright, who introduced and investigated it between 1933 and 1940 [163–167]. We note that originally Wright considered such a function restricted to $\lambda \geq 0$ in his paper [165]

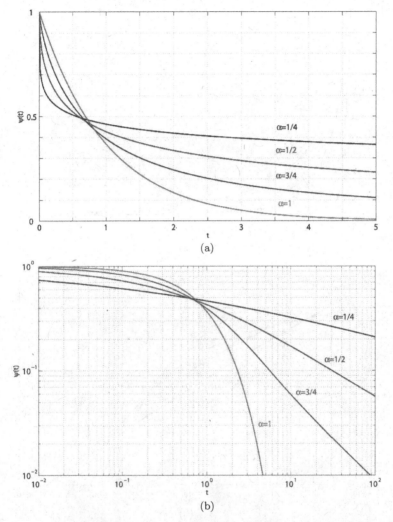

Figure 3.1. Plots of $\psi_\alpha(t)$ with $\alpha = 1/4, 1/2, 3/4, 1$ versus t; left: linear scales $(0 \leq t \leq 5)$; right: logarithmic scales $(10^{-2} \leq t \leq 10^2)$.

in connection with his investigations in the asymptotic theory of partitions. Only later, in 1940, he extended to $-1 < \lambda < 0$ [166].

Like for the Mittag-Leffler functions, a description of the most important properties of the Wright functions (with relevant references up to the fifties) can be found in the third volume of the

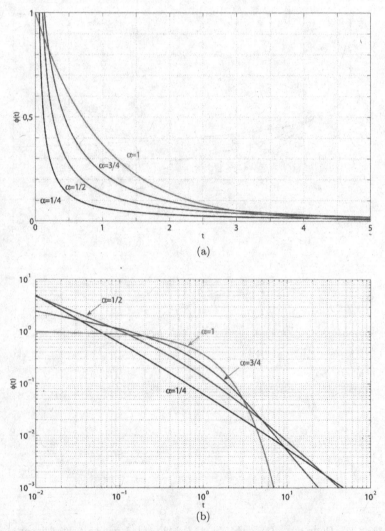

Figure 3.2. Plots of $\phi_\alpha(t)$ with $\alpha = 1/4, 1/2, 3/4, 1$ versus t; left: linear scales ($0 \leq t \leq 5$); right: logarithmic scales ($10^{-2} \leq t \leq 10^2$).

Bateman Project [42], in the chapter *XVIII* on the *Miscellaneous Functions*. However, probably for a misprint, there λ is restricted to be positive.

Relevant investigations on the Wright functions have been carried out by Stanković [150], among other authors quoted in Kiryakova's book [77, p. 336], and, more recently, by Luchko and Gorenflo [86], Gorenflo, Luchko and Mainardi [52, 53] and Luchko [85].

The special cases $\lambda = -\nu$, $\mu = 0$ and $\lambda = -\nu$, $\mu = 1 - \nu$ with $0 < \nu < 1$ and z replaced by $-z$ provide the Wright-type functions, $F_\nu(z)$ and $M_\nu(z)$, respectively, that have been so denoted and investigated by Mainardi [89–94]. Since these functions are of special interest for us, we shall later return to them and to present a detailed analysis, see also [52, 53]. We refer to them as the auxiliary functions of the Wright type.

3.7.1. The series representation of the Wright function

The Wright function is defined by the power series convergent in the whole complex plane,

$$\Phi_{\lambda,\mu}(z) := \sum_{n=0}^{\infty} \frac{z^n}{n!\,\Gamma(\lambda n + \mu)}, \quad \lambda > -1, \quad \mu \in \mathbb{C}, \quad z \in \mathbb{C}. \quad (3.7.1)$$

The case $\lambda = 0$ is trivial since the Wright function is reduced to the exponential function with the constant factor $1/\Gamma(\mu)$, which turns out to vanish identically for $\mu = -n$, $n = 0, 1, \ldots$. In general it is proved that the Wright function for $\lambda > -1$ and $\mu \in \mathbb{C}$ ($\mu \neq -n$, $n = 0, 1, \ldots$ if $\lambda = 0$) is an entire function of finite order ρ and type σ given by, see e.g., [52].

$$\rho = \frac{1}{1 + \lambda}, \quad \sigma = (1 + \lambda) \, |\lambda|^{\lambda/(1+\lambda)}. \quad (3.7.2)$$

In particular, the Wright function turns out to be of *exponential type* if $\lambda \geq 0$.

3.7.2. The Wright integral representation and asymptotic expansions

The integral representation of the Wright function reads

$$\Phi_{\lambda,\mu}(z) = \frac{1}{2\pi i} \int_{Ha} e^{\zeta + z\zeta^{-\lambda}} \frac{d\zeta}{\zeta^\mu}, \quad \lambda > -1, \quad \mu \in \mathbb{C}, \quad z \in \mathbb{C}. \quad (3.7.3)$$

Here Ha denotes an arbitrary Hankel path, namely a contour consisting of pieces of the two rays $\arg \zeta = \pm\phi$ extending to infinity, and of the circular arc $\zeta = \epsilon\, e^{i\theta}$, $|\theta| \leq \phi$, with $\phi \in (\pi/2, \pi)$, and $\epsilon > 0$, arbitrary. The identity between the integral and series representations is obtained by using the Hankel representation of the reciprocal of the gamma function.

The complete picture of the asymptotic behavior of the Wright function for large values of z was given by Wright himself by using the method of steepest descent on the integral representation (3.7.3). In particular the papers [164, 165] were devoted to the case $\lambda > 0$ and the paper [166] to the case $-1 < \lambda < 0$. Wright's results have been summarized by Gorenflo, Luchko and Mainardi [52, 53]. Recently, Wong and Zhao [161, 162] have provided a detailed asymptotic analysis of the Wright function in the cases $\lambda > 0$ and $-1 < \lambda < 0$ respectively, achieving a uniform "exponentially improved" expansion with a smooth transition across the Stokes lines. The asymptotics of zeros of the Wright function has been investigated by Luchko [85].

Here we limit ourselves to recall Wright's result [167] in the case $\lambda = -\nu \in (-1, 0)$, $\mu > 0$, where the following asymptotic expansion is valid in a suitable sector about the negative real axis,

$$\Phi_{-\nu,\mu}(z) = Y^{1/2-\mu}\, e^{-Y} \left(\sum_{m=0}^{M-1} A_m Y^{-m} + O(|Y|^{-M}) \right), \quad |z| \to \infty,$$

(3.7.4)

with $Y = (1 - \nu)(-\nu^\nu z)^{1/(1-\nu)}$, where the A_m are certain real numbers.

3.7.3. The Wright functions as generalization of the Bessel functions

For $\lambda = 1$ and $\mu = \nu + 1$ the Wright function turns out to be related to the well known Bessel functions J_ν and I_ν by the following identity:

$$(z/2)^\nu \Phi_{1,\nu+1}(\mp z^2/4) = \begin{cases} J_\nu(z), \\ I_\nu(z). \end{cases}$$

(3.7.5)

In view of this property some authors refer to the Wright function as the *Wright generalized Bessel function* (misnamed also as the *Bessel–Maitland function*) and introduce the notation

$$J_\nu^{(\lambda)}(z) := \left(\frac{z}{2}\right)^\nu \sum_{n=0}^\infty \frac{(-1)^n (z/2)^{2n}}{n!\,\Gamma(\lambda n + \nu + 1)}; \quad J_\nu^{(1)}(z) := J_\nu(z). \quad (3.7.6)$$

As a matter of fact, the Wright function appears as the natural generalization of the entire function known as *Bessel–Clifford function*, see, e.g., Kiryakova [77, p. 336], and referred by Tricomi, see, e.g., [49, 152] as the *uniform Bessel function*

$$T_\nu(z) := z^{-\nu/2} J_\nu(2\sqrt{z}) = \sum_{n=0}^\infty \frac{(-1)^n z^n}{n!\,\Gamma(n + \nu + 1)} = \Phi_{1,\nu+1}(-z).$$
$$(3.7.7)$$

Some of the properties which the Wright functions share with the popular Bessel functions were enumerated by Wright himself. Hereafter, we quote two relevant relations from the Bateman Project [42], which can easily be derived from (3.7.1) or (3.7.3):

$$\lambda z\, \Phi_{\lambda,\lambda+\mu}(z) = \Phi_{\lambda,\mu-1}(z) + (1 - \mu)\, \Phi_{\lambda,\mu}(z), \quad (3.7.8)$$

$$\frac{d}{dz} \Phi_{\lambda,\mu}(z) = \Phi_{\lambda,\lambda+\mu}(z). \quad (3.7.9)$$

3.7.4. The auxiliary functions $F_\nu(z)$ and $M_\nu(z)$ of the Wright function

In our treatment of the time fractional diffusion wave equation we find it convenient to introduce two *auxiliary functions* $F_\nu(z)$ and $M_\nu(z)$, where z is a complex variable and ν a real parameter $0 < \nu < 1$. Both functions turn out to be analytic in the whole complex plane, i.e., they are entire functions. Their respective *integral representations* read,

$$F_\nu(z) := \frac{1}{2\pi i} \int_{Ha} e^{\zeta - z\zeta^\nu}\, d\zeta, \quad 0 < \nu < 1, \ z \in \mathbb{C}, \quad (3.7.10)$$

$$M_\nu(z) := \frac{1}{2\pi i} \int_{Ha} e^{\zeta - z\zeta^\nu}\, \frac{d\zeta}{\zeta^{1-\nu}}, \quad 0 < \nu < 1, \ z \in \mathbb{C}. \quad (3.7.11)$$

From a comparison of (3.7.10)–(3.7.11) with (3.7.3) we easily recognize that

$$F_\nu(z) = \Phi_{-\nu,0}(-z), \qquad (3.7.12)$$

and

$$M_\nu(z) = \Phi_{-\nu,1-\nu}(-z). \qquad (3.7.13)$$

From (3.7.8) and (3.7.12)–(3.7.13) we find the relation

$$F_\nu(z) = \nu\, z\, M_\nu(z). \qquad (3.7.14)$$

This relation can be obtained directly from (3.7.10)–(3.7.11) via an integration by parts,

$$\int_{Ha} e^{\zeta - z\zeta^\nu}\, \frac{\mathrm{d}\zeta}{\zeta^{1-\nu}} = \int_{Ha} e^\zeta \left(-\frac{1}{\nu z}\frac{\mathrm{d}}{\mathrm{d}\zeta}e^{-z\zeta^\nu}\right)\mathrm{d}\zeta = \frac{1}{\nu z}\int_{Ha} e^{\zeta - z\zeta^\nu}\,\mathrm{d}\zeta.$$

The *series representations* for our auxiliary functions turn out to be, respectively,

$$F_\nu(z) := \sum_{n=1}^\infty \frac{(-z)^n}{n!\,\Gamma(-\beta n)}$$

$$= -\frac{1}{\pi}\sum_{n=1}^\infty \frac{(-z)^n}{n!}\,\Gamma(\nu n + 1)\sin(\pi\nu n), \qquad (3.7.15)$$

$$M_\nu(z) := \sum_{n=0}^\infty \frac{(-z)^n}{n!\,\Gamma[-\nu n + (1-\nu)]}$$

$$= \frac{1}{\pi}\sum_{n=1}^\infty \frac{(-z)^{n-1}}{(n-1)!}\,\Gamma(\nu n)\sin(\pi\nu n). \qquad (3.7.16)$$

The series at the right-hand side have been obtained by using the well-known reflection formula for the Gamma function $\Gamma(\zeta)\,\Gamma(1-\zeta) = \pi/\sin\pi\zeta$. Furthermore, we note that $F_\nu(0) = 0$, $M_\nu(0) = 1/\Gamma(1-\nu)$ and that the relation (3.7.14) can be derived also from (3.7.15)–(3.7.16).

Explicit expressions of $F_\nu(z)$ and $M_\nu(z)$ in terms of known functions are expected for some particular values of ν. Mainardi and

Tomirotti [110] have shown that for $\nu = 1/q$, with an integer $q \geq 2$, the auxiliary function $M_\nu(z)$ can be expressed as a sum of $(q-1)$ simpler entire functions, namely,

$$M_{1/q}(z) = \frac{1}{\pi} \sum_{h=1}^{q-1} c(h, q)\, G(z; h, q) \qquad (3.7.17)$$

with

$$c(h, q) = (-1)^{h-1}\, \Gamma(h/q)\, \sin(\pi h/q), \qquad (3.7.18)$$

$$G(z; h, q) = \sum_{m=0}^{\infty} (-1)^{m(q+1)} \left(\frac{h}{q}\right)_m \frac{z^{qm+h-1}}{(qm+h-1)!}. \qquad (3.7.19)$$

Here $(a)_m$, $m = 0, 1, 2, \ldots$, denotes Pochhammer's symbol

$$(a)_m := \frac{\Gamma(a+m)}{\Gamma(a)} = a(a+1)\ldots(a+m-1).$$

We note that $(-1)^{m(q+1)}$ is equal to $(-1)^m$ for q even and $+1$ for q odd. In the particular cases $q = 2$, $q = 3$ we find:

$$M_{1/2}(z) = \frac{1}{\sqrt{\pi}} \exp(-z^2/4), \qquad (3.7.20)$$

$$M_{1/3}(z) = 3^{2/3}\, \mathrm{Ai}(z/3^{1/3}), \qquad (3.7.21)$$

where Ai denotes the *Airy function*, see, e.g., [42].

Furthermore, it can be proved that $M_{1/q}(z)$ (for integer ≥ 2) satisfies the differential equation of order $q - 1$,

$$\frac{d^{q-1}}{dz^{q-1}} M_{1/q}(z) + \frac{(-1)^q}{q} z\, M_{1/q}(z) = 0, \qquad (3.7.22)$$

subjected to the $q - 1$ initial conditions at $z = 0$, derived from the series expansion in (3.7.17)–(3.7.19),

$$M_{1/q}^{(h)}(0) = \frac{(-1)^h}{\pi}\, \Gamma[(h+1)/q]\, \sin[\pi\,(h+1)/q],$$

$$h = 0, 1, \ldots, q - 2. \qquad (3.7.23)$$

We note that, for $q \geq 4$, Eq. (3.7.22) is akin to the *hyper-Airy* differential equation of order $q - 1$, see, e.g., [14]. Consequently, the function $M_\nu(z)$ is a generalization of the hyper-Airy function. In the limiting case $\nu = 1$ we get $M_1(z) = \delta(z - 1)$, i.e., the M-function degenerates into a generalized function of Dirac type.

From our purposes (time-fractional diffusion processes) it is relevant to consider the M_ν-function for a positive (real) argument that will be denoted by r. Later, by using its Laplace transform with the Bernstein theorem, we shall prove that $M_\nu(r) > 0$ for $r > 0$.

The asymptotic representation of $M_\nu(r)$, as $r \to \infty$ can be obtained by using the ordinary saddle-point method. Choosing as a variable r/ν rather than r the computation is easier and yields, see [110],

$$M_\nu(r/\nu) \sim a(\nu)\, r^{(\nu-1/2)/(1-\nu)} \exp[-b(\nu)\, r^{(1/(1-\nu))}], \quad r \to +\infty, \tag{3.7.24}$$

where $a(\nu) = 1/\sqrt{2\pi\,(1-\nu)} > 0$, and $b(\nu) = (1-\nu)/\nu > 0$.

The above asymptotic representation is consistent with the first term of the asymptotic expansion (3.7.4) obtained by Wright for $\Phi_{-\nu,\mu}(-r)$. In fact, taking $\mu = 1 - \nu$ so $1/2 - \mu = \nu - 1/2$, we obtain

$$M_\nu(r) \sim A_0\, Y^{\nu-1/2} \exp(-Y), \quad r \to \infty, \tag{3.7.25}$$

where

$$A_0 = \frac{1}{\sqrt{2\pi}\,(1-\nu)^\nu\, \nu^{2\nu-1}}, \quad Y = (1-\nu)\,(\nu^\nu\, r)^{1/(1-\nu)}. \tag{3.7.26}$$

Because of the above exponential decay, any moment of order $\delta > -1$ for $M_\nu(r)$ is finite. In fact,

$$\int_0^\infty r^\delta\, M_\nu(r)\, \mathrm{d}r = \frac{\Gamma(\delta+1)}{\Gamma(\nu\delta+1)}, \quad \delta > -1,\ 0 < \nu < 1. \tag{3.7.27}$$

In particular we get the normalization property in $\mathbb{R}^+$, $\int_0^\infty M_\nu(r)\, \mathrm{d}r = 1$.

Similarly, we can compute any moment of order $\delta > -1$ of the generic function $\Phi_{-\nu,\mu}(-r)$ in view of its exponential decay (3.7.4), obtaining

$$\int_0^\infty r^\delta \, \Phi_{-\nu,\mu}(-r) \, dr$$

$$= \frac{\Gamma(\delta+1)}{\Gamma(\nu\delta+\nu+\mu)}, \quad \delta > -1, \quad 0 < \nu < 1, \quad \mu > 0. \quad (3.7.28)$$

We also quote an interesting formula derived by Stankovič [150], which provides a relation between the Whittaker function $W_{-1/2,1/6}$ and the Wright function $\Phi_{-2/3,0} = F_{2/3}$,

$$F_{2/3}(x^{-2/3}) = -\frac{1}{2\sqrt{3\pi}}\exp\left(-\frac{2}{27x^2}\right) W_{-1/2,1/6}\left(-\frac{4}{27x}\right).$$

$$(3.7.29)$$

We recall that the generic Whittaker function $W_{\lambda,\mu}(x)$ satisfies the differential equation, see, e.g., [42]

$$\frac{d^2}{dx^2}W_{\lambda,\mu}(x) + \left(-\frac{1}{4} + \frac{\lambda}{x} + \frac{\mu^2}{4x^2}\right) W_{\lambda,\mu}(x) = 0, \quad \lambda, \mu \in \mathbb{R}.$$

$$(3.7.30)$$

3.7.5. Laplace transform pairs related to the Wright function

Let us now consider some Laplace transform pairs related to the Wright functions. We continue to denote by r a positive variable.

In the case $\lambda > 0$ the Wright function is an entire function of order less than 1 and consequently, being of exponential type, its Laplace transform can be obtained by transforming term-by-term its Taylor expansion (3.7.1) in the origin, see, e.g., [41]. As a result we get

$$\Phi_{\lambda,\mu}(\pm r) \overset{\mathcal{L}}{\leftrightarrow} \frac{1}{s}\sum_{k=0}^\infty \frac{(\pm s^{-1})^k}{\Gamma(\lambda k + \mu)} = \frac{1}{s}E_{\lambda,\mu}(\pm s^{-1}), \quad \lambda > 0, \quad \mu \in \mathbb{C}.$$

$$(3.7.31)$$

Here $E_{\alpha,\beta}(z)$ denotes the generalized Mittag-Leffler function (3.6.5). In this case the resulting Laplace transform turns out to be analytic for $s \neq 0$, vanishing at infinity and exhibiting an essential singularity at $s = 0$.

For $-1 < \lambda < 0$ the just applied method cannot be used since then the Wright function is an entire function of order greater than one. In this case, setting $\nu = -\lambda$, the existence of the Laplace transform of the function $\Phi_{-\nu,\mu}(-t)$, $t > 0$, follows from (3.7.4), which says us that the function $\Phi_{-\nu,\mu}(z)$ is exponentially small for large z in a sector of the plane containing the negative real semi-axis. To get the transform in this case we can use the idea given in [94] based on the integral representation (3.7.3). We have

$$\Phi_{-\nu,\mu}(-r) \overset{\mathcal{L}}{\leftrightarrow} \frac{1}{2\pi i} \int_{Ha} \frac{e^{\zeta}\zeta^{-\mu}}{s + \zeta^{\nu}} \, d\zeta = E_{\nu,\mu+\nu}(-s), \qquad (3.7.32)$$

where we have used the integral representation (3.6.10) of the generalized Mittag-Leffler function.

The relation (3.7.32) was given in [39] (see also [38]) in the case $\mu \geq 0$ as a representation of the generalized Mittag-Leffler function in the whole complex plane as a Laplace integral of an entire function but without identifying this function as the Wright function. They also gave (in slightly different notations) the more general representation

$$E_{\alpha_2,\beta_2}(z) = \int_0^{\infty} E_{\alpha_1,\beta_1}(z r^{\alpha_1}) r^{\beta_1 - 1} \Phi_{-\nu,\gamma}(-r) \, dr, \qquad (3.7.33)$$

with $0 < \alpha_2 < \alpha_1$, β_1, $\beta_2 > 0$, and $0 < \nu = -\alpha_2/\alpha_1 < 1$, $\gamma = \beta_2 - \beta_1 \alpha_2/\alpha_1$.

An important particular case of the Laplace transform pair (3.7.32) is given for $\mu \doteq 1 - \nu$ to yield, see also [94],

$$M_{\nu}(r) \overset{\mathcal{L}}{\leftrightarrow} E_{\nu}(-s), \quad 0 < \nu < 1. \qquad (3.7.34)$$

As a further particular case we recover the well-known Laplace transform pair, see, e.g., [41],

$$M_{1/2}(r) = \frac{1}{\sqrt{\pi}} \exp(-r^2/4) \overset{\mathcal{L}}{\leftrightarrow} E_{1/2}(-s) := \exp(s^2) \operatorname{erfc}(s).$$
$$\qquad (3.7.35)$$

We also note that, transforming term-by-term the Taylor series of $M_{\nu}(r)$ (not being of exponential order) yields a series of negative powers of s, which represents the asymptotic expansion of $E_\nu(-s)$ as $s \to \infty$ in a sector around the positive real semi-axis.

Using the relation

$$\int_0^\infty r^n f(r) \, dr = \lim_{s \to 0} (-1)^n \frac{d^n}{ds^n} \mathcal{L}\{f(r); s\},$$

the Laplace transform pair (3.7.32) and the series representation of the generalized Mittag-Leffler function. (3.6.5) we can compute all the moments of integer order for the Wright function $\Phi_{-\nu,\mu}(-r)$ with $0 < \nu < 1$ in $\mathbb{R}^+$:

$$\int_0^\infty r^n \Phi_{-\nu,\mu}(-r) \, dr = \frac{n!}{\Gamma(\nu n + \mu + \nu)}, \quad n \in \mathbb{N}_0. \tag{3.7.36}$$

This formula is consistent with the more general formula (3.7.28) valid when the moments are of arbitrary order $\delta > -1$.

We can now obtain other Laplace transform pairs related to our auxiliary functions. Indeed, following [94] and using the integral representations (3.7.10)–(3.7.11) we get

$$\frac{1}{r} F_\nu(cr^{-\nu}) = \frac{c\nu}{r^{\nu+1}} M_\nu(cr^{-\nu}) \overset{\mathcal{L}}{\leftrightarrow} \exp(-cs^\nu), \quad 0 < \nu < 1, \ c > 0. \tag{3.7.37}$$

The Laplace inversion in Eq. (3.7.37) was properly carried out by Pollard [134, 135] (based on a formal result by Humbert [69]) and by Mikusiński [121]. A formal series inversion was carried out by Buchen and Mainardi [18], albeit unaware of the previous results.

By applying the formula for differentiation of the image of the Laplace transform to Eq. (3.7.37), we get a Laplace transform pair useful for our further discussions, namely

$$\frac{1}{r^\nu} M_\nu(cr^{-\nu}) \overset{\mathcal{L}}{\leftrightarrow} s^{\nu-1} \exp(-cs^\nu), \quad 0 < \nu < 1, \ c > 0. \tag{3.7.38}$$

As particular cases of Eqs. (3.7.37)–(3.7.38), we recover the well-known pairs, see, e.g., [41],

$$\frac{1}{2r^{3/2}} M_{1/2}(1/r^{1/2})) = \frac{1}{2\sqrt{\pi}} r^{-3/2} \exp(-1/(4r^2)) \overset{\mathcal{L}}{\leftrightarrow} \exp(-s^{1/2}), \tag{3.7.39}$$

$$\frac{1}{r^{1/2}}M_{1/2}(1/r^{1/2})) = \frac{1}{\sqrt{\pi}}\,r^{-1/2}\exp(-1/(4r^2)) \overset{\mathcal{L}}{\leftrightarrow} s^{-1/2}\exp(-s^{1/2}).$$
$$(3.7.40)$$

More generally, using the same method as in (3.7.37), we get (see [150]), the pair

$$r^{\mu-1}\Phi_{-\nu,\mu}(-cr^{-\nu}) \overset{\mathcal{L}}{\leftrightarrow} s^{-\mu}\exp(-cs^{\nu}), \quad 0 < \nu < 1, \ c > 0.$$
$$(3.7.41)$$

Stanković [150] also gave some other pairs related to the Wright function including:

$$r^{\mu/2-1}\,\Phi_{-\nu,\mu}(-r^{-\nu/2}) \overset{\mathcal{L}}{\leftrightarrow} \frac{\sqrt{\pi}}{2^{\mu}}\,s^{-\mu/2}\,\Phi_{-\nu/2,(\mu+1)/2}(-2^{\nu}s^{\nu/2}),$$
$$(3.7.42)$$

with $0 < \nu < 1$, and

$$r^{-\mu}\exp(-r^{\nu}\cos(\nu\pi))\sin(\mu\pi + r^{\nu}\sin(\nu\pi)) \div \pi s^{\mu-1}\Phi_{-\nu,\mu}(-s^{-\nu}),$$
$$(3.7.43)$$

with $0 < \nu < 1$ and $\mu < 1$.

3.7.6. Some plots of the M-Wright functions

For readers' convenience we find it instructive to show the plots of the M-Wright functions on the real axis for some rational values of the parameter ν.

In Figs. 3.3 and 3.4 we compare the plots of these functions in $-5 \le x \le 5$ for some rational values in the ranges $\nu \in [0, 1/2]$ and $\nu \in [1/2, 1]$, respectively. To gain more insight of the effect of the parameter itself on the behavior close to and far from the origin, we have adopted both linear and logarithmic scale for the ordinates. Consequently, in Fig. 3.3 we see the transition from $\exp(-|x|)$ for $\nu = 0$ to $1/\sqrt{\pi}\exp(-x^2)$ for $\nu = 1/2$, whereas in Fig. 3.4 we see the transition from $1/\sqrt{\pi}\exp(-x^2)$ to the delta function $\delta(1 - |x|)$ for $\nu = 1$.

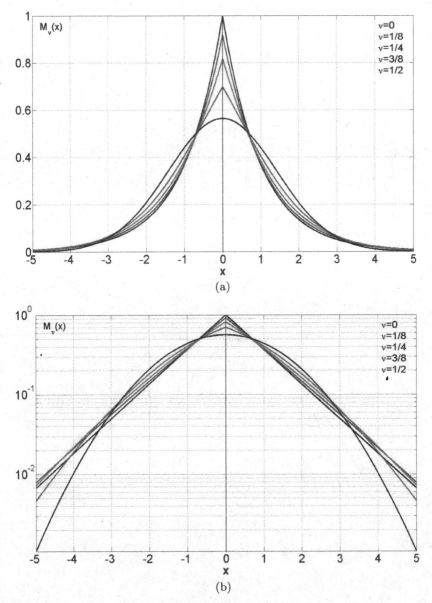

Figure 3.3. Plots of the Wright-type function $M_\nu(x)$ with $\nu = 0, 1/8, 1/4, 3/8,$ $1/2$ for $-5 \le x \le 5$; left: linear scale, right: logarithmic scale.

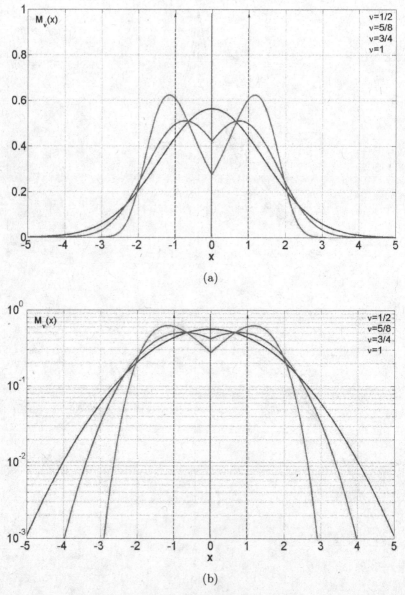

(a)

(b)

Figure 3.4.   Plots of the Wright-type function $M_\nu(x)$ with $\nu = 1/2,\ 5/8,\ 3/4,\ 1$ for $-5 \leq x \leq 5$: left: linear scale; right: logarithmic scale.

Bibliography

[1]   N.H. Abel, Solution de quelques problèmes à l'aide d'intégrales définie [Nor-wegian], *Magazin for Naturvidenskaberne*, Aargang 1, Bind 2, Christiana 1823. French translation in *Oeuvres Complètes*, Vol. I, pp. 11–18. Nouvelle èdition par L. Sylow et S. Lie, 1881.

[2]   N.H. Abel, Aufloesung einer mechanischen Aufgabe, *J. Reine Angew. Math. (Crelle)*, **I** (1826), 153–157.

[3]   M. Abramowitz and I.A. Stegun, *Handbook of Mathematical Functions*, Dover, New York, 1965.

[4]   R.P. Agarwal, A propos d'une note de M. Pierre Humbert, *C.R. Acad. Sci. Paris* **236** (1953), 2031–2032.

[5]   T.M. Atanacković, S. Pilipović, B. Stanković and D. Zorica, *Fractional Calcuklus with Applications in Mechanics*, 2 Vols., Wiley, New York, 2014.

[6]   Yu.I. Babenko, *Heat and Mass Transfer*, Chimia, Leningrad, 1986 [in Russian].

[7]   R.T. Baillie and M.L. King (Editors), Fractional Differencing and Long Memory Processes, *J. Econom.* **73**(1) (1996), 1–324.

[8]   A.V. Balakrishnan, Representation of abstract Riesz potentials of the elliptic type, *Bull. Amer. Math. Soc.* **64**(5) (1958), 266–289.

[9]   A.V. Balakrishnan, Operational calculus for infinitesimal generators of semi-groups, *Trans. Amer. Math. Soc.* **91**(2) (1959), 330–353.

[10] A.V. Balakrishnan, Fractional powers of closed operators and the semi-groups generated by them, *Pacific J. Math.* **10**(2) (1960), 419–437.

[11] D. Baleanu, K. Diethelm, E. Scalas and J.J. Trujillo, *Fractional Calculus: Models and Numerical Methods*, World Scientific, Singapore (2012).

[12] R. Balescu, V-Langevin equations, continuous time random walks and fractional diffusion, *Chaos, Solitons Fractals* **34** (2007), 62–80.

[13] J.H. Barret, Differential equations of non-integer order, *Canad. J. Math.* **6** (1954), 529–541.

[14] C.M. Bender and S.A. Orszag, *Advanced Mathematical Methods for Scientists and Engineers*, McGraw-Hill, Singapore, 1987, Chapter 3.

[15] L. Blank, Numerical treatment of differential equations of fractional order, MCCM Numerical Analysis Report No. 287, The University of Manchester, 1996. [www.ma.man.ac.uk/MCCM/MCCM.html].

[16] S. Bochner, Diffusion equation and stochastic processes, *Proc. Natl. Acad. Sci. USA* **35** (1949), 368–370.

[17] B.L.J. Braaksma, Asymptotic expansions and analytical continuations for a class of Barnes-integrals, *Compositio Math.* **15** (1962), 239–341.

[18] P.W. Buchen and F. Mainardi, Asymptotic expansions for transient viscoelastic waves, *J. Mécanique* **14** (1975), 597–608.

[19] P. Butzer and U. Westphal, Introduction to fractional calculus, in: H. Hilfer (ed.), *Fractional Calculus, Applications in Physics*, World Scientific, Singapore, 2000, pp. 1–85.

[20] E. Capelas de Oliveira, F. Mainardi and J. Vaz Jr, Models based on Mittag-Leffler functions for anomalous relaxation in dielectrics,

European Phys. J. Special Topics **193** (2011), 161–171. E-print http://arxiv.org/abs/1106.1761, [Revised Version].

[21] M. Caputo, Linear models of dissipation whose Q is almost frequency independent: Part II., *Geophys. J. R. Astr. Soc.* **13** (1967), 529–539.

[22] M. Caputo, *Elasticità e Dissipazione*, Zanichelli, Bologna, 1969 [in Italian].

[23] M. Caputo, Vibrations of an infinite viscoelastic layer with a dissipative memory, *J. Acoust. Soc. Am.* **56**(3) (1974), 897–904.

[24] M. Caputo, Which is the correct stress strain relation for the anelasticity of the earth's interior? *Geophys. J. R. Astr. Soc.* **58** (1979), 432–435.

[25] M. Caputo, Generalized rheology and geophysical consequences, *Tectonophysics* **116** (1985), 163–172.

[26] M. Caputo, The rheology of an anelastic medium studied by means of the observation of the splitting of its eigenfrequencies, *J. Acoust. Soc. Am.* **86**(5) (1989), 1984–1989.

[27] M. Caputo, Modern rheology and electric induction: multivalued index of refraction, splitting of eigenvalues and fatigues, *Ann. Geofisica* **39**(5) (1996), 941–966.

[28] M. Caputo, Models of flux in porous media with memory, *Water Resources Res.* **36**(3) (2000), 693–705.

[29] M. Caputo, Distributed order differential equations modelling dielectric induction and diffusion, *Fract. Calc. Appl. Anal.* **4** (2001), 421–444.

[30] M. Caputo and F. Mainardi, A new dissipation model based on memory mechanism, *Pure Appl. Geophys. (PAGEOPH)* **91** (1971), 134–147.

[31] M. Caputo and F. Mainardi, Linear models of dissipation in anelastic solids, *Riv. Nuovo Cimento (Ser. II)* **1** (1971), 161–198.

[32] A. Carpinteri and F. Mainardi (eds.), *Fractals and Fractional Calculus in Continuum Mechanics*, Springer, Wien, 1997.

[33] K.S. Cole, Electrical conductance of biological systems, in: *Proc. Symp. Quant. Biol.*, Cold Spring Harbor, New York, 1933, pp. 107–116.

[34] V. Daftardar-Gejji (ed.), *Fractional Calculus, Theory and Applications*, Narosa Publishing House, New Delhi, 2014.

[35] H.T. Davis, *The Theory of Linear Operators*, The Principia Press, Bloomington, IN, 1936.

[36] K. Diethelm, *The Analysis of Fractional Differential Equations*, Lecture Notes in Mathematics, Springer, Berlin, 2010.

[37] M.M. Djrbashian, *Integral Transforms and Representations of Functions in the Complex Plane*, Nauka, Moscow, 1966 [in Russian].

[38] M.M. Djrbashian, *Harmonic Analysis and Boundary Value Problems in the Complex Domain*, Birkhäuser, Basel, 1993.

[39] M. M. Djrbashian and R. A. Bagian, On integral representations and measures associated with Mittag-Leffler type functions, *Izv. Akad. Nauk Armjanskvy SSR, Matematika* **10** (1975), 483–508 [in Russian].

[40] M.M. Djrbashian and A.B. Nersesian, Fractional derivatives and the Cauchy problem for differential equations of fractional order, *Izv. Acad. Nauk Armjanskvy SSR, Matematika* **3**(1) (1968), 3–29. [in Russian]

[41] G. Doetsch, *Introduction to the Theory and Applications of the Laplace Transformation*, Springer, Berlin, 1974.

[42] A. Erdélyi (ed.), *Higher Transcendental Functions*, Bateman Project, Vols. 1–3, McGraw-Hill, New York, 1953–1955.

[43] A. Erdélyi (ed.), *Tables of Integral Transforms*, Bateman Project, Vols. 1–2, McGraw-Hill, New York, 1953–1955.

[44] W. Feller, On a generalization of Marcel Riesz' potentials and the semi-groups generated by them, *Meddelanden Lunds Universitets Matematiska Seminarium* (Comm. Sém. Mathém. Université de Lund), Tome suppl. dédié a M. Riesz, Lund (1952), pp. 73–81.

[45] W. Feller, *An Introduction to Probability Theory and its Applications*, Vol. 2, 2nd edn. Wiley, New York, 1971 [1st edn. 1966].

[46] C. Fox, The G and H functions as symmetrical Fourier kernels, *Trans. Amer. Math. Soc.* **98** (1961), 395–429.

[47] Lj. Gajić and B. Stanković, Some properties of Wright's function, *Publ. de l'Institut Mathèmatique, Beograd, Nouvelle Sèr.* **20**(34) (1976), 91–98.

[48] R. Garra, A. Giusti, F. Mainardi and G. Pagnini, Fractional relaxation with time varying coefficient, *Fract. Calc. Appl. Anal.* **17**(2) (2014), 424–439.

[49] L. Gatteschi, *Funzioni Speciali*, UTET, Torino, 1973, pp. 196–197.

[50] I.M. Gel'fand and G.E. Shilov, *Generalized Functions*, Vol. 1, Academic Press, New York, 1964 [English translation from the Russian (Nauka, Moscow, 1959)].

[51] R. Gorenflo, Fractional calculus: some numerical methods, in: A. Carpinteri and F. Mainardi (eds.), *Fractals and Fractional Calculus in Continuum Mechanics*, Springer Verlag, Wien, 1997, pp. 277–290 [Reprinted in www.fracalmo.org].

[52] R. Gorenflo, Yu. Luchko and F. Mainardi, Analytical properties and applications of the Wright function, *Fract. Calc. Appl. Anal.* **2**(4) (1999), 383–414.

[53] R. Gorenflo, Yu. Luchko and F. Mainardi, Wright functions as scale-invariant solutions of the diffusion-wave equation, *J. Comput. Appl. Math.* **118**(1–2) (2000), 175–191.

[54] R. Gorenflo and F. Mainardi, Fractional oscillations and Mittag-Leffler functions, Preprint A-14/96, Fachbereich Mathematik und Informatik, Freie Universität, Berlin, 1996 [www.math.fu-berlin.de/publ/index.html].

[55] R. Gorenflo and F. Mainardi, Fractional calculus: integral and differential equations of fractional order, in: A. Carpinteri and F. Mainardi (eds.), *Fractals and Fractional Calculus in Continuum Mechanics*, Springer, Wien, 1997, pp. 223–276 [Reprinted in www.fracalmo.org].

[56] R. Gorenflo and F. Mainardi, Random walk models for space-fractional diffusion processes, *Fract. Calc. Appl. Anal.* **1**(2) (1998), 167–191.

[57] R. Gorenflo and F. Mainardi, Approximation of Lévy-Feller diffusion by random walk, *J. Anal. Appl. (ZAA)* **18**(2) (1999), 231–246.

[58] R. Gorenflo and F. Mainardi, Random walk models approximating symmetric space-fractional diffusion processes, in: J. Elschner, I. Gohberg

and B. Silbermann (eds.), *Problems in Mathematical Physics* (Siegfried Prössdorf Memorial Volume), Birkhäuser, Boston, 2001, pp. 120–145.

[59] R. Gorenflo and S. Vessella, *Abel Integral Equations: Analysis and Applications*, Springer-Verlag, Berlin, 1991 [Lecture Notes in Mathematics, Vol. 1461].

[60] B. Gross, On creep and relaxation, *J. Appl. Phys.* **18** (1947), 212–221.

[61] A.K. Grünwald, Über "begrenzte" Derivation und deren Anwendung, *Z. Angew. Math. Phys.* **12** (1867), 441–480.

[62] K.G. Gupta and U.C. Jain, The H-function: II, *Proc. Natl. Acad. Sci. India A* **36** (1966), 594–602.

[63] H. Haubold, A.M. Mathai and R.K. Saxena, Mittag-Leffler Functions and Their Applications, *J. Appl. Math.*, **2011** (2011), Article ID 298628, 51 pp.

[64] R. Hilfer, Fractional Derivatives in Static and Dynamic Scaling, in B. Dubrulle, F. Graner and D. Sornette (eds.), *Scale Invariance and Beyond*, Springer Verlag, Berlin and EDP Science, France, 1977, pp. 53–62. Lecture 3, Les Houches Workshop, March 10–14, 1997.

[65] R. Hilfer (ed.), *Applications of Fractional Calculus in Physics*, World Scientific, Singapore 2000.

[66] R. Hilfer and H.J. Seybold, Computation of the generalized Mittag-Leffler function and its inverse in the complex plane, *Integral Transforms Spec. Funct.* **17**(9) (2006), 637–652.

[67] E. Hille and J.D. Tamarkin, On the theory of linear integral equations *Ann. of Math.* **31** (1930), 479–528.

[68] Hj. Holmgren, Om differentialkalkylen med indices af hvad natur som helst, *Kongl. Svenska Vetenskaps-Akad. Hanl. Stockholm* **5**(11) (1865–1866), 1–83.

[69] P. Humbert, Nouvelles correspondances symboliques, *Bull. Sci. Mathém. (Paris, II Ser.)* **69** (1945), 121–129.

[70] P. Humbert, Quelques résultats relatifs à la fonction de Mittag-Leffler, *C.R. Acad. Sci. Paris* **236** (1953), 1467–1468.

[71] P. Humbert and R.P. Agarwal, Sur la fonction de Mittag-Leffler et quelques-unes de ses généralisations, *Bull. Sci. Math (Ser. II)* **77** (1953), 180–185.

[72] A.A. Kilbas and M. Saigo, On Mittag-Leffler type functions, fractional calculus operators and solution of integral equations, *Integral Transforms Spec. Funct.* **4** (1996), 355–370.

[73] A.A. Kilbas and M. Saigo, On the H functions, *J. Appl. Math. Stoch. Anal.* **12** (1999), 191–204.

[74] A.A. Kilbas and M. Saigo, *H-transforms: Theory and Applications*, CRC Press, 2004.

[75] A.A. Kilbas, M. Saigo and J.J. Trujillo, On the generalized Wright function, *Fract. Calc. Appl. Anal.* **5** (2002), 437–460.

[76] A.A. Kilbas, H.M. Srivastava and J.J. Trujillo, *Theory and Applications of Fractional Differential Equations*, Elsevier, Amsterdam, 2006 [North-Holland Mathematics Studies, No. 204].

[77] V. Kiryakova, *Generalized Fractional Calculus and Applications*, Longman, Harlow 1994 [Pitman Research Notes in Mathematics, #301].

[78] A.N. Kochubei, A Cauchy problem for evolution equations of fractional order, *Differential Equations* **25** (1989), 967–974 [English translation from the Russian Journal *Differentsial'nye Uravneniya*].

[79] A.N. Kochubei, Fractional order diffusion, *Differential Equations* **26** (1990), 485–492 [English translation from the Russian Journal *Differentsial'nye Uravneniya*].

[80] A.V. Letnikov, Theory of differentiation with an arbitrary index, *Mat. Sb.* **3** (1868), 1–66 [in Russian].

[81] A.V. Letnikov, On historical development of differentiation theory with an arbitrary index, *Mat. Sb.* **3** (1868), 85–112 [in Russian].

[82] J. Liouville, Mémoire sur quelques questions de géométrie et de mécanique, et sur un nouveau genre de calcul pour résoudre ces questions, *J. École Roy. Polytéchn.* **13**(Sect. 21) (1832), 1–69.

[83] J. Liouville, Mémoire sur le calcul des différentielles à indices quelconques, *J. École Roy. Polytéchn.* **13**(Sect. 21) (1832), 71–162.

[84] J. Liouville, Mémoire sur l'intégration des équations différentielles à indices fractionnaires, *J. École Roy. Polytéchn.* **15**(55) (1837), 55–84.

[85] Yu. Luchko, Asymptotics of zeros of the Wright function, *J. Anal. Appl. (ZAA)* **19** (2000), 583–596.

[86] Yu. Luchko and R. Gorenflo, Scale-invariant solutions of a partial differential equation of fractional order, *Fract. Calc. Appl. Anal.* **1** (1998), 63–78.

[87] Yu. Luchko, F. Mainardi, and Yu. Povstenko, Propagation speed of the maximum of the fundamental solution to the fractional diffusion-wave equation, *Comput. Math. Appl.* **66** (2013), 774–784 [E-print: http://arxiv.org/abs/1201.5313].

[88] R.L. Magin, *Fractional Calculus in Bioengineering*, Begell House Publishers, Connecticut, 2006.

[89] F. Mainardi, On the initial value problem for the fractional diffusion-wave equation, in: S. Rionero and T. Ruggeri (eds.), *Waves and Stability in Continuous Media*, World Scientific, Singapore, 1994, pp. 246–251.

[90] F. Mainardi, Fractional diffusive waves in viscoelastic solids, in: J.L. Wegner and F.R. Norwood (eds.), *Nonlinear Waves in Solids*, ASME book No. AMR 137, Fairfield NJ 1995, pp. 93–97 [*Abstr. Appl. Mech. Rev.* **46**(12) (1993), 549].

[91] F. Mainardi, The time fractional diffusion-wave equation, *Radiofisika* **38**(1–2) (1995), 20–36 [English Transl.: *Radiophysics & Quantum Electronics*].

[92] F. Mainardi, The fundamental solutions for the fractional diffusion-wave equation, *Appl. Math. Lett.* **9**(6) (1996), 23–28.

[93] F. Mainardi, Fractional relaxation-oscillation and fractional diffusion-wave phenomena, *Chaos, Solitons Fractals* **7** (1996), 1461–1477.

[94] F. Mainardi, Fractional calculus: some basic problems in continuum and statistical mechanics, in: A. Carpinteri and F. Mainardi (eds.), *Fractals*

and Fractional Calculus in Continuum Mechanics, Springer, Wien, 1997, pp. 291–348 [E-print: http://arxiv.org/abs/1201.0863].

[95] F. Mainardi, Applications of integral transforms in fractional diffusion processes, *Integral Transforms Special Functions* **15**(6) (2004), 477–484 [E-print: http://arxiv.org/abs//0710.0145].

[96] F. Mainardi, *Fractional Calculus and Waves in Linear Viscoelasticity*, Imperial College Press, World Scientific, London and Singapore, 2010.

[97] F. Mainardi and R. Gorenflo, The Mittag-Leffler function in the Riemann–Liouville fractional calculus, in: A.A. Kilbas (ed.), *Boundary Value Problems, Special Functions and Fractional Calculus*, Belarusian State University, Minsk, 1996, pp. 215–225.

[98] F. Mainardi and R. Gorenflo, On Mittag-Leffler type functions in fractional evolution processes, *J. Comput. Appl. Math.* **118** (2000), 283–299.

[99] F. Mainardi and R. Gorenflo, Fractional calculus: special functions and applications, in: G. Dattoli, H.M. Srivastava and D. Cocolicchio (eds.), *Advanced Special Functions and Applications*, Aracne, Roma, 2000, pp. 165–189.

[100] F. Mainardi and R. Gorenflo, Time-fractional derivatives in relaxation processes: a tutorial survey, *Fract. Calc. Appl. Anal.* **10** (2007), 269–308 [E-print: http://arxiv.org/abs/0801.4914].

[101] F. Mainardi, R. Gorenflo and A. Vivoli, Renewal processes of Mittag-Leffler and Wright type, *Fractional Calculus Appl. Anal.* **8**(1) (2005), 7–38 [E-print: http://arxiv.org/abs/math/0701455].

[102] F. Mainardi, Yu. Luchko and G. Pagnini, The fundamental solution of the space-time fractional diffusion equation, *Fractional Calculus Appl. Anal.* **4**(2) (2001), 153–192 [E-print: http://arxiv.org/abs/cond-mat/0702419].

[103] F. Mainardi, A. Mura, R. Gorenflo and M. Stojanovic, The two forms of fractional relaxation of distributed order, *J. Vibration Control* **13**(9–10) (2007), 1249–1268 [E-print: http://arxiv.org/abs/cond-mat/0701131].

[104] F. Mainardi, A. Mura, G. Pagnini and R. Gorenflo, Time-fractional diffusion of distributed order, *J. Vibration Control* **14**(9–10) (2008), 1267–1290 [arxiv.org/abs/cond-mat/0701132].

[105] F. Mainardi, A. Mura and G. Pagnini, The *M*-Wright function in time-fractional diffusion processes: a tutorial survey, *Int. J. Differential Equations* **2010** (2010), Article ID 104505, 29 pp. [E-print: http://arxiv.org/abs/1004.2950].

[106] F. Mainardi and G. Pagnini, The Wright functions as solutions of the time-fractional diffusion equations, *Appl. Math. Comput.* **141**(1) (2003), 51–66.

[107] F. Mainardi and G. Pagnini, Salvatore Pincherle: the pioneer of the Mellin-Barnes integrals, *J. Comput. Appl. Math.* **153** (2003), 331–342 [E-print: http://arxiv.org/abs/math/0702520].

[108] F. Mainardi, G. Pagnini and R. Gorenflo, Mellin transform and subordination laws in fractional diffusion processes, *Fractional Calculus Appl. Anal.* **6**(4) (2003), 441–459 [E-print: http://arxiv.org/abs/math/0702133].

[109] F. Mainardi, G. Pagnini and R.K. Saxena, Fox H functions in fractional diffusion, *J. Comput. Appl. Math.* **178**(1-2) (2005), 321-331.

[110] F. Mainardi and M. Tomirotti, On a special function arising in the time fractional diffusion-wave equation, in: P. Rusev, I. Dimovski and V. Kiryakova (eds.), *Transform Methods and Special Functions*, Sofia, 1994, Science Culture Technology, Singapore, 1995, pp. 171-183.

[111] F. Mainardi and M. Tomirotti, Seismic pulse propagation with constant Q and stable probability distributions, *Ann. Geofisica* **40** (1997), 1311-1328.

[112] A. Marchaud, Sur les dérivées et sur les différences des fonctions de vriables réelles, *J. Math. Pures Appl.* **6**(4) (1927), 337-425.

[113] A.M. Mathai, *Handbook of Generalized Special Functions in Statistics and Physical Sciences*, Clarendon Press, Oxford, 1993.

[114] A.M Mathai and H.J. Haubold, *Special Functions for Applied Scientists*, Springer Science, New York, 2008.

[115] A.M. Mathai and R.K. Saxena, *Generalized Hypergeometric Function with Applications in Statistics and Physical Sciences*, Springer, Berlin, 1973 [Lecture Notes in Mathematics, No. 348].

[116] A.M. Mathai and R.K. Saxena, *The H-function with Applications in Statistics and Other Disciplines*, Wiley Eastern Ltd, New Delhi, 1978.

[117] A.M. Mathai, R.K. Saxena and H.J. Haubold, A certain class of Laplace transforms with applications to reaction and reaction-diffusion equations, *Astrophys. Space Sci.* **305** (2006), 283-288.

[118] A.M. Mathai, R.K. Saxena and H.J. Haubold, *The H-function: Theory and Applications*, Springer, New York, 2010.

[119] A.C. McBride, *Fractional Calculus and Integral Transforms of Generalized Functions*, Pitman, London 1979 [Pitman Research Notes in Mathematics, #31].

[120] A.C. McBride and G.F. Roach (eds.), *Fractional Calculus*, Pitmand, London, 1985 [Pitman Research Notes in Mathematics, #138].

[121] J. Mikusiński, On the function whose Laplace transform is $\exp(-s^\alpha \lambda)$, *Studia Math.* **18** (1959), 191-198.

[122] K.S. Miller, The Weyl fractional calculus, in B. Ross (ed.), *Fractional Calculus and its Applications*, Springer, Berlin, 1975, pp. 80-89 [Lecture Notes in Mathematics, #457].

[123] K.S. Miller and B. Ross, *An Introduction to the Fractional Calculus and Fractional Differential Equations*, Wiley, New York, 1993.

[124] M.G. Mittag-Leffler, Sur l'intégrale de Laplace-Abel, *C.R. Acad. Sci. Paris* (*Ser. II*) **136** (1902), 937-939.

[125] M.G. Mittag-Leffler, Une généralisation de l'intégrale de Laplace-Abel, *C.R. Acad. Sci. Paris* (*Ser. II*) **137** (1903), 537-539.

[126] M.G. Mittag-Leffler, Sur la nouvelle fonction $E_\alpha(x)$, *C.R. Acad. Sci. Paris* (*Ser. II*) **137** (1903), 554-558.

[127] M.G. Mittag-Leffler, Sopra la funzione $E_\alpha(x)$, *R. Accad. Lincei, Rend.* (*Ser. V*) **13** (1904), 3-5.

[128] M.G. Mittag-Leffler, Sur la représentation analytique d'une branche uniforme d'une fonction monogène, *Acta Math.* **29** (1905), 101–181.

[129] K. Nishimoto (ed.), *Fractional Calculus and its Applications*, Nihon University, Tokyo, 1990. [Proc. Int. Conf. held at Nihon Univ., Tokyo (Japan), 1989].

[130] K. Nishimoto, *An Essence of Nishimoto's Fractional Calculus*, Descartes Press, Koriyama, 1991.

[131] K.B. Oldham and J. Spanier, *The Fractional Calculus*, Academic Press, New York, 1974.

[132] S. Pincherle, Sulle funzioni ipergeometriche generalizzate, *Atti R. Accademia Lincei, Rend. Cl. Sci. Fis. Mat. Nat. (Ser. 4)* **4** (1888), 694–700, 792–799 [Reprinted in *Salvatore Pincherle: Opere Scelte*, edited by UMI (Unione Matematica Italiana) Vol. 1, pp. 223–230, 231–239, Cremonese, Roma, 1954].

[133] I. Podlubny, *Fractional Differential Equations*, Academic Press, San Diego, 1999.

[134] H. Pollard, The representation of $\exp(-x^\lambda)$ as a Laplace integral, *Bull. Amer. Math. Soc.* **52** (1946), 908–910.

[135] H. Pollard, The completely monotonic character of the Mittag-Leffler function $E_\alpha(-x)$, *Bull. Amer. Math. Soc.* **54** (1948), 1115–1116.

[136] B. Riemann, Versuch einer allgemeinen Auffassung der Integration und Differentiation, in: *Bernhard Riemann's gesammelte mathematische Werke und wissenschaftliker Nachlass*, pp. 331–344, Teubner, Leipzig, 1876 [New edition edited by H. Weber, Dover, New York, 1953].

[137] M. Riesz, L'intégrales de Riemann–Liouville et le probléme de Cauchy, *Acta Math.* **81**(1–2) (1949), 1–223.

[138] B. Ross (Editor), *Fractional Calculus and its Applications*, Springer, Berlin, 1975 [Lecture Notes in Mathematics, #457].

[139] B. Rubin, *Fractional Integrals and Potentials*, Addison-Wesley and Longman, Harlow 1996 [Pitman Monographs and Surveys in Pure and Applied Mathematics, #82].

[140] P. Rusev, I. Dimovski and V. Kiryakova (eds.), *Transform Methods and Special Functions*, Sofia, 1994, Science Culture Technology, Singapore, 1995 [Proc. 1st Int. Workshop TMSF, Sofia, Bulgaria, 12–17 August 1994].

[141] P. Rusev, I. Dimovski, and V. Kiryakova (eds.), *Transform Methods and Special Functions, Varna 1996*, Inst. Maths & Informatics, Bulg. Acad. Sci, Sofia 1998 [Proc. 2nd Int. Workshop TMSF, Varna, Bulgaria, 23–29 August 1996].

[142] A. Saichev and G. Zaslavsky, Fractional kinetic equations: solutions and applications, *Chaos* **7** (1997), 753–764.

[143] S.G. Samko, A.A. Kilbas and O.I. Marichev, *Fractional Integrals and Derivatives, Theory and Applications*, Gordon and Breach, Amsterdam, 1993. [English translation from the Russian, *Integrals and Derivatives of Fractional Order and Some of Their Applications*, Nauka i Tekhnika, Minsk, 1987].

[144] G. Sansone and J. Gerretsen, *Lectures on the Theory of Functions of a Complex Variable, Vol. I. Holomorphic Functions*, Nordhoff, Groningen, 1960, pp. 345-349.

[145] K. Sato, *Lévy Processes and Infinitely Divisible Distributions*, Cambridge University Press, Cambridge, 1999.

[146] W.R. Schneider, Stable distributions: Fox function representation and generalization, in: S. Albeverio, G. Casati and D. Merlini (eds.), *Stochastic Processes in Classical and Quantum Systems*, Springer, Berlin, pp. 497-511 [Lecture Notes in Physics, Vol. 262].

[147] H.J. Seybold and R. Hilfer, Numerical results for the generalized Mittag-Leffler function, *Fract. Calc. Appl. Anal.* **8** (2005), 127-139.

[148] H.J. Seybold and R. Hilfer, Numerical algorithm for calculating the generalized Mittag-Leffler function, *SIAM J. Numer. Anal.* **47**(1) (2008), 69-88.

[149] H.M. Srivastava, K.C. Gupta and S.P. Goyal, *The H-functions of One and Two Variables with Applications*, South Asian Publishers, New Delhi, 1982.

[150] B. Stankovič, On the function of E.M. Wright, *Publ. Inst, Math. Beograd, Nouvelle Sèr.* **10** (1970), 113-124.

[151] H. Takayasu, *Fractals in the Physical Sciences*, Manchester University Press, Manchester and New York, 1990.

[152] F.G. Tricomi, *Fonctions Hypergéometriques Confluentes*, Gauthier-Villars, Paris, 1960 [Mém. Sci. Math., #140].

[153] V.V. Uchaikin, *Fractional Derivatives for Physicists and Engineers*, 2 Vols. Springer, Berlin — Higher Education Press, Beijing, 2013.

[154] V.V. Uchaikin and V.M. Zolotarev, *Change and Stability. Stable Distributions and their Applications*, VSP, Utrecht, 1999 [Series "Modern Probability and Statistics", No. 3].

[155] Vu Kim Tuan and R. Gorenflo, Extrapolation to the limit for numerical fractional differentiation, *ZAMM* **75** (1995), 646-648.

[156] B.J. West, M. Bologna and P. Grigolini, *Physics of Fractal Operators*, Springer Verlag, New York, 2003 [Institute for Nonlinear Science].

[157] U. Westphal, An approach to fractional powers of operators via fractional differences, *Proc. London Math. Soc.* **29**(3) (1974), 557-576.

[158] U. Westphal, Fractional powers of infinitesimal generators of semigroups, in: R. Hilfer (ed.), *Applications of Fractional Calculus in Physics*, World Scientific, Singapore 2000, pp. 131-170.

[159] H. Weyl, Bemerkungen zum Begriff des Differentialquotientten gebrochener Ordnung, *Vierteljahresschr. Naturforsch. Ges. Zürich* **62**(1–2) (1917), 296-302.

[160] A. Wiman, Über die Nullstellen der Funktionen $E_\alpha[X]$, *Acta Math.* **29** (1905), 217-234.

[161] R. Wong and Y.-Q. Zhao, Smoothing of Stokes' discontinuity for the generalized Bessel function, *Proc. R. Soc. London A* **455** (1999), 1381-1400.

[162] R. Wong and Y.-Q. Zhao, Smoothing of Stokes' discontinuity for the generalized Bessel function II, *Proc. R. Soc. London A* **455** (1999), 3065-3084.

[163] E.M. Wright, On the coefficients of power series having exponential singularities, *J. London Math. Soc.* **8** (1933), 71–79.

[164] E.M. Wright, The asymptotic expansion of the generalized Bessel function, *Proc. London Math. Soc. (Ser. II)* **38** (1935), 257–270.

[165] E.M. Wright, The asymptotic expansion of the generalized hypergeometric function, *J. London Math. Soc.* **10** (1935), 287–293.

[166] E.M. Wright, The asymptotic expansion of the generalized Bessel function, *Proc. London Math. Soc. (Ser. II)* **46** (1940), 389–408.

[167] E.M. Wright, The generalized Bessel function of order greater than one, *Quart. J. Math., Oxford Ser.* **11** (1940), 36–48.

[168] G.M. Zaslavsky, *Hamiltonian Chaos and Fractional Dynamics*, Oxford University Press, Oxford, 2005.

Chapter 4

Fractional Calculus and Fractional Differential Equations*

4.1. Introduction

Fractional calculus (FC) deals with differentiation and integration of arbitrary orders. In 1695 l'Hôpital in a letter to Leibniz, asked what is the meaning of $\frac{d^n y}{dx^n}$ when $n = \frac{1}{2}$? Since then this topic has developed through the pioneering works of Euler (1730), Lagrange (1772), Laplace (1812), Fourier (1822), Liouville (1834), Riemann (1847) and many other mathematicians of 18th and 19th century. The first application of FC was given by Abel (1823) for solving the tautochrone problem. Subsequently operational methods proposed by Heaviside (1850–1925) have been instrumental in solving the engineering problems. FC has history of more than three hundred years. The upsurge of research activity in this area has revived since 1974 with the appearance of the classic book by Oldham and Spanier [1], as the potential applications of FC to various branches of Science and Engineering became apparent. Numerous problems in Physics, Chemistry, Engineering and Biological Sciences are better described in terms of Fractional Differential Equations (FDEs) [2, 3]. FC is a new tool which has widened the descriptive power of calculus beyond the familiar integer order concepts of rates of change and area under a curve. The purpose of these lectures is to introduce the

---

*This chapter is based on the lectures of Professor Varsha Daftardar-Gejji, Department of Mathematics, Savitribai Phule Pune University, Pune-411007, India.

concepts of fractional calculus and differential equations of fractional order.

The lectures have been organized as follows. In Section 4.2, we begin with the concept of differ–integral, and derive its properties. Further we define Riemann–Liouville fractional integral and derivative, Caputo derivative and related results in terms of their inter-relations, composition rules, Laplace transforms and so on. We illustrate how to solve linear FDE using Laplace transform method. Further we introduce Fractional Differential Equations (FDEs) and state existence, uniqueness and stability of their solutions. In Section 4.3, we focus on the methods for solving linear/nonlinear FDEs. We explain Adomian Decomposition Method (ADM) and New Iterative Method (NIM), which have widely been used in the recent literature. This is followed by illustrative examples demonstrating the utility of these methods. Extension of separation of variables method for Fractional Boundary Value Problems (FBVPs) is detailed out in Section 4.4. Section 4.5 deals with systems of FDEs. We begin with linear systems and the results on the existence of solutions and stability analysis. Further we explain the methods to solve them using the basics of Linear Algebra. Further we turn our attention to nonlinear systems with special emphasis on chaos theory and give a brief introduction to chaos in nonlinear FDEs. Finally, we present recent results including stable manifold theorem for nonlinear fractional systems.

4.2. Fractional Derivatives/Integrals

4.2.1. Differ–integral

Grünwald (1890) developed the notion of fractional derivative/ integral as the limit of a difference quotient. The ordinary derivatives are defined in terms of backward differences as

$$
\begin{aligned}
\frac{\mathrm{d}f}{\mathrm{d}x} &= \lim_{h \to 0} \frac{[f(x) - f(x-h)]}{h}, \\
\frac{\mathrm{d}^2 f}{\mathrm{d}x^2} &= \lim_{h \to 0} \frac{[f(x) - 2f(x-h) + f(x-2h)]}{h^2}.
\end{aligned}
\tag{4.2.1}
$$

In general, for $n \in N$, and $f \in C^n[a, b]$, $a < x < b$,

$$D^n f(x) = \lim_{h \to 0} \frac{\sum_{j=0}^{n} [(-1)^j \binom{n}{j} f(x - jh)]}{h^n}. \qquad (4.2.2)$$

Let us first see how to modify (4.2.2) so that it will represent n-fold integral for negative integral values of n. Consider base point a and define for $x > a$, $h_N = \frac{x-a}{N}$, N is a positive integer and the symbol $\binom{n}{j}$ can be generalized for negative integers as

$$\binom{-n}{j} := \frac{-n(-n-1)\cdots(-n-j+1)}{j!}.$$

Replacing n by $-n$ in (4.2.2), we get

$$D_a^{-n} f(x) = \lim_{N \to \infty} h_N^n \left[\sum_{j=0}^{N} (-1)^j \binom{-n}{j} f(x - jh_N) \right]. \qquad (4.2.3)$$

Exercise 4.2.1. Show that

$$D_a^{-n} f(x) = \int_a^x \frac{(x - \tau)^{n-1}}{(n - 1)!} f(\tau) d\tau. \qquad (4.2.4)$$

Exercise 4.2.2. Show that

$$\int_a^x \frac{(x - \tau)^{n-1}}{(n - 1)!} f(\tau) d\tau = \underbrace{\int_a^x dx_1 \int_a^{x_1} dx_2 \cdots \int_a^{x_{n-1}} f(t) dt}_{n \text{ times}}. \qquad (4.2.5)$$

Grünwald–Letnikov definition is an extension of the formula (4.2.2) to arbitrary order $q \in \mathbb{R}$.

Definition 4.2.1. Grünwald–Letnikov differ–integral of arbitrary order q is defined as

$$\mathbb{D}_a^q f(x) = \lim_{N \to \infty} h_N^{-q} \left[\sum_{j=0}^{N} (-1)^j \binom{q}{j} f(x - jh_N) \right], \qquad q \in R, \qquad (4.2.6)$$

where

$$\binom{q}{0} = 1, \quad \binom{q}{j} = \frac{q(q-1)\ldots(q-j+1)}{j!}, \quad j \in \mathbb{N}.$$

Lemma 4.2.1 ([1]). *For $q \in R$, and $n \in N$*

$$\frac{d^n}{dx^n} \mathbb{D}_a^q f(x) = \mathbb{D}_a^{n+q} f(x).$$

Proof.

$$\frac{d}{dx} \mathbb{D}_a^q f(x) = \lim_{N \to \infty} \frac{\mathbb{D}_a^q f(x) - \mathbb{D}_a^q f(x - h_N)}{h_N}, \tag{4.2.7}$$

as $N \to \infty$, $h_N \to 0$.

$$\mathbb{D}_a^q f(x) = \lim_{N \to \infty} h_N^{-q} \left\{ \sum_{j=0}^{N} (-1)^j \binom{q}{j} f(x - jh_N) \right\}$$

$$= \lim_{N \to \infty} h_N^{-q} \left\{ f(x) + \sum_{j=1}^{N} (-1)^j \binom{q}{j} f(x - jh_N) \right\}, \tag{4.2.8}$$

$$\mathbb{D}_a^q f(x - h_N) = \lim_{N \to \infty} h_N^{-q} \left\{ \sum_{j=0}^{N-1} (-1)^j \binom{q}{j} f(x - jh_N - h_N) \right\}$$

$$= \lim_{N \to \infty} h_N^{-q} \left\{ \sum_{j=1}^{N} (-1)^{j-1} \binom{q}{j-1} f(x - jh_N) \right\}. \tag{4.2.9}$$

Substituting (4.2.8) and (4.2.9) in (4.2.7), we get

$$\frac{d}{dx} \mathbb{D}_a^q f(x)$$

$$= \lim_{N \to \infty} h_N^{-q-1} \left\{ f(x) + \sum_{j=1}^{N} (-1)^j \left[\binom{q}{j} + \binom{q}{j-1} \right] f(x - jh_N) \right\}$$

$$= \lim_{N \to \infty} h_N^{-q-1} \left\{ \sum_{j=0}^{N} (-1)^j \binom{q+1}{j} f(x - jh_N) \right\}$$

$$= \mathbb{D}_a^{q+1} f(x). \tag{4.2.10}$$

Applying the operator $\frac{\mathrm{d}}{\mathrm{d}x}$, n-times, we arrive at the result.

4.2.2. Riemann–Liouville fractional integral/derivative

Riemann–Liouville (RL) definition of fractional integral (Liouville, 1832, Riemann, 1847) is a generalization Cauchy's formula (4.2.4), in which n-fold integral is replaced by a non-integer value q. RL definition existed prior to GL differ-integral formula. If $f(x) \in C[a, b]$ and $q > 0$ then the (left-sided) Riemann–Liouville fractional integral I_a^q is defined as

$$I_a^q f(x) := \frac{1}{\Gamma(q)} \int_a^x \frac{f(\tau)}{(x - \tau)^{1-q}} \mathrm{d}\tau, \quad x > a. \tag{4.2.11}$$

Riemann–Liouville fractional derivative of order q, $n-1 \leq q < n$, $n \in N$, is defined as

$$D_a^q f(x) = \frac{\mathrm{d}^n}{\mathrm{d}x^n} \left[\int_a^x \frac{(x - \tau)^{-q+n-1}}{\Gamma(n - q)} f(\tau) \mathrm{d}\tau \right]. \tag{4.2.12}$$

Repeated, integration by parts of (4.2.12) (n-times) yields

$$D_a^q f(x) = \sum_{k=0}^{n-1} \frac{f^{(k)}(a)(x - a)^{k-q}}{\Gamma(k - q + 1)}$$

$$+ \frac{1}{\Gamma(n - q)} \int_a^x (x - \tau)^{n-q-1} f^{(n)}(\tau) \mathrm{d}\tau, \tag{4.2.13}$$

where $n - 1 \leq q < n$, $f \in C^n$.

Remarks.

- For $f \in C[a, b]$, and $q > 0$, $\mathbb{D}_a^{-q} f(x)$ turns out to be equal to $I_a^q f(x)$. For $n - 1 < q \leq n$, and $f \in C^n$, $\mathbb{D}_a^q f(x) = D_a^q f(x)$.
- GL definition is convenient for numerical calculations.

Composition rules

(1) $D^n D_a^q = D_a^{n+q}$, $n \in N$, $q > 0$.

(2)
$$D^n I_a^q = \begin{cases} D_a^{n-q} & \text{if } n > q, \\ I_a^{q-n} & \text{if } n < q. \end{cases}$$

(3) $\qquad\qquad I_a^p \, I_a^q f(x) = I_a^{p+q} f(x), \quad p, q \geq 0.$

(4) Let $p > 0$, $m - 1 \leq q < m$, then
$$D_a^q I_a^p = \begin{cases} I_a^{-q+p} & \text{if } p \geq q, \\ D_a^{q-p} & \text{if } p \leq q. \end{cases}$$

In particular for $p > 0$,
$$D_a^p I_a^p f(x) = f(x).$$

(5) $I_a^p \, D_a^p f(x)$
$$= f(x) - \sum_{j=0}^{n-1} \frac{D_a^{p-j} f(a)(x-a)^{p-j}}{\Gamma(p-j+1)}, \quad n - 1 \leq p < n, \; n \in N.$$

Exercise 4.2.3. Prove the composition rules.

Exercise 4.2.4. Show that

$$I_a^q (x-a)^p = \frac{\Gamma(p+1)}{\Gamma(p+q+1)} (x-a)^{p+q}, \quad p > -1, \; q \geq 0, \quad (4.2.14)$$

$$D_a^q (x-a)^p = \frac{\Gamma(p+1)}{\Gamma(p-q+1)} (x-a)^{p-q}, \quad p > -1, \; q \geq 0. \quad (4.2.15)$$

We enlist below some important properties, proofs of which can be found in [1].

(1) If the infinite series of function $\sum_{k=0}^{\infty} f_k$ converges uniformly in $0 < |x - a| < l$, then

$$I_a^\alpha \left(\sum_{k=0}^{\infty} f_k(x) \right) = \sum_{k=0}^{\infty} I_a^\alpha f_k(x), \quad \alpha \geq 0, \qquad (4.2.16)$$

and right-hand side also converges uniformly in $0 < |x - a| < l$.

(2) If the infinite series $\sum_{k=0}^{\infty} f_k$ as well as the series $\sum_{k=0}^{\infty} D_a^\alpha f_k(x)$ converge uniformly in $0 < |x - a| < l$, then

$$D_a^\alpha \left(\sum_{k=0}^{\infty} f_k(x) \right) = \sum_{k=0}^{\infty} D_a^\alpha f_k(x), \quad \alpha \geq 0, \quad (4.2.17)$$

for $0 < |x - a| < l$.

(3) The Leibniz rule for fractional derivatives takes the following form:

$$D_a^q(fg)(x) = \sum_{j=0}^{\infty} \binom{q}{j} D_a^{q-j} f(x) g^{(j)}(x). \quad (4.2.18)$$

Laplace transform

The Laplace transform of a function $f(x), x > 0$, is defined as

$$\mathcal{L}[f(x)](s) = F(s) = \int_0^{\infty} e^{-sx} f(x) \mathrm{d}x, \quad s \in \mathbb{C}. \quad (4.2.19)$$

If the integral (4.2.19) is convergent at a point $s_0 \in \mathbb{C}$, then it converges absolutely for $s \in \mathbb{C}$, such that $\Re(s) > \Re(s_0)$. The infimum σ of all values $\Re(s)$ for which the Laplace integral (4.2.19) converges is called as abscissa of convergence.

The inverse Laplace transform is given for $x > 0$ by the formula,

$$\mathcal{L}^{-1}[f(s)](x)$$
$$= \frac{1}{2\pi i} \int_{y-i\infty}^{y+i\infty} e^{sx} F(s) \mathrm{d}s, \quad s = x + iy \in \mathbb{C}, \quad \Re(s) > \sigma.$$

$$(4.2.20)$$

For "sufficiently nice" functions:

$$\mathcal{L}^{-1}\mathcal{L}f = f, \quad \mathcal{L}\mathcal{L}^{-1}F = F.$$

We enlist some important theorems pertaining to Laplace transform.

Theorem 4.2.1. *If* $f(x) = \sum_{n=0}^{\infty} a_n x^n$ *converges for* $x \geq 0$, *and* $|a_n| \leq K \frac{\lambda^n}{n!}, \forall n$ *where* $\lambda > 0, K > 0$, *then*

$$\mathcal{L}[f(x)](s) = \sum_{n=0}^{\infty} a_n \mathcal{L}[x^n](s) = \sum_{n=0}^{\infty} \frac{a_n n!}{s^{n+1}}, \quad \Re(s) > \lambda.$$

Theorem 4.2.2. *If* $\mathcal{L}[f(x)](s) = F(s)$, *then*

(i) $F(s-a) = \mathcal{L}[e^{ax}f(x)](s)$, $a \in \mathbb{R}$, $\Re(s) > 0$;

(ii) $\frac{d^n}{ds^n}F(s) = \mathcal{L}[(-1)^n x^n f(x)](s)$, $s > \sigma$, $n \in \mathbb{N}$.

Theorem 4.2.3. *Suppose* f *is continuous on* $(0, \infty)$ *and of exponential order* α *and* f' *is piecewise continuous on* $[0, \infty)$. *Then*

$$\mathcal{L}[f'(x)](s) = sF(s) - f(0^+), \quad \Re(s) > \sigma.$$

More generally, suppose $f(x), \dots, f^{(n-1)}(x)$ *are continuous on* $[0, \infty]$, *then*

$$\mathcal{L}[f^{(n)}(x)](s) = s^n F(s) - s^{n-1} f(0^+)$$
$$- s^{n-2} f'(0^+) - \cdots - f^{(n-1)}(0^+).$$

Theorem 4.2.4 (Laplace Convolution Theorem). *If* f *and* g *are piecewise continuous on* $[0, \infty)$ *and of exponential order* α, *then*

$$\mathcal{L}[(f * g)(x)](s) = F(s)G(s), \quad \Re(s) > \sigma,$$

where $[(f * g)(x)] = \int_0^x f(\tau)g(x - \tau)\mathrm{d}\tau$.

Exercise 4.2.5. Calculate the following:

(1) $\mathcal{L}[x^n e^{ax}](s)$;

(2) $\mathcal{L}[e^{2x} \sin(3x)](s)$;

(3) $\mathcal{L}[\sinh \omega x](s)$;

(4) $\mathcal{L}[x \cos \omega x](s)$;

(5) $\mathcal{L}^{-1}\left[\log \frac{s+a}{s+b}\right](x)$;

(6) $\mathcal{L}^{-1}\left[\frac{s}{s^2+4s+1}\right](x)$.

Riemann–Liouville fractional integral

$$I_0^q f(x) = \frac{1}{\Gamma(q)} \int_0^x (x-t)^{q-1} f(t)\mathrm{d}t = \frac{x^{q-1}}{\Gamma(q)} * f(x), \qquad (4.2.21)$$

where $*$ denotes Laplace convolution. Hence using Laplace convolution theorem

$$\mathcal{L}[I_0^q f(x), s] = s^{-q} F(s), \quad \Re(s) > 0. \tag{4.2.22}$$

For finding the Laplace transform of RL derivative of order q, $n - 1 \leq q < n$, we note

$$D_0^q f(x) = g^{(n)}(x), \quad \text{where } g(x) = I_0^{n-q} f(x),$$

$$\mathcal{L}[D_0^q f(x), s] = \mathcal{L}[g^{(n)}(x), s] = s^n G(s) - \sum_{k=0}^{n-1} s^k g^{(n-k-1)}(0).$$

Hence in view of (4.2.22)

$$\mathcal{L}[D_0^q f(x), s] = s^q F(s) - \sum_{k=0}^{n-1} s^k D_0^{q-k-1} f(0), \quad \Re(s) > 0. \tag{4.2.23}$$

Some comments are in order. In view of the expression (4.2.23) for the Laplace transform, the practical applicability of RL derivative is limited as the physical interpretation of the terms $D_0^{q-k-1} f(0)$ does not exist. From the view point of applications, the fractional derivative introduced by Caputo [4] is receiving more attention.

4.2.3. Caputo derivative

Definition 4.2.2. Caputo derivative of order p, $n - 1 < p < n$, $n \in N$, is defined for $f \in C^n$, $x > a$, as

$$^c D_a^p f(x) = \frac{1}{\Gamma(n-p)} \int_a^x (x-\tau)^{n-p-1} f^{(n)}(\tau) d\tau = \mathcal{I}_a^{n-p} D^n f(x),$$

$$^c D_a^n f(x) = f^{(n)}(x) \quad \text{for } n \in N.$$

Laplace transform of Caputo derivative

$$\mathcal{L}\{^c D_0^p f(x), s\} = \int_0^\infty e^{-st} \{^c D_0^p f(t)\} dt$$

$$= s^p F(s) - \sum_{k=0}^{n-1} s^{p-k-1} f^{(k)}(0), \quad \Re(s) > 0. \tag{4.2.24}$$

Remark. Initial conditions required are of the form $f^{(k)}(0)$, which are physically meaningful.

Relation between RL and Caputo derivative

Let $n - 1 \leq q < n$, $n \in N$,

$$D_a^q f(x) = D^n \left(I_a^{n-q} f(x) \right) = D^n \left(I_a^{n-q} \{ D^n I_a^n f \} \right)$$

$$= D^n \left(I_a^{n-q} \left\{ I_a^n f^{(n)} + \sum_{k=0}^{n-1} f^k(a) \frac{(x-a)^k}{k!} \right\} \right)$$

$$= I_a^{n-q} f^{(n)} + \sum_{k=0}^{n-1} f^k(a) \frac{(x-a)^{k-q}}{\Gamma(k+1-q)}. \tag{4.2.25}$$

Hence

$$D_a^q f(x) = {}^c D_a^q f(x) + \sum_{k=0}^{n-1} \frac{f^{(k)}(a)(x-a)^{k-q}}{\Gamma(k-q+1)}.$$

Thus, $D_a^q = {}^c D_a^q$ if and only if $f^{(k)}(a) = 0$, $0 \leq k \leq n-1$.

Composition rules

Let $f \in C^n[a, b]$ and $n - 1 < q < n$ then

$$I_a^q \, {}^c D_a^q f(x) = I_a^q I_a^{n-q} f^{(n)}(x) = I^n f^{(n)}(x)$$

$$= f(x) - \sum_{k=0}^{n-1} \frac{f^{(k)}(a)(x-a)^k}{k!}, \tag{4.2.26}$$

$${}^c D_a^q I_a^q f(x) = f(x). \tag{4.2.27}$$

Leibniz rule

In view of the relation (4.2.25) and the Leibniz rule for RL derivative given in (4.2.18), it is easy to verify that if f and g are C^∞-functions, for $n - 1 < q < n$,

$$^cD_a^q(f(x)g(x)) = \sum_{k=0}^{\infty} \binom{q}{k} D_a^{q-k} f(x) \, g^{(k)}(x)$$

$$- \sum_{k=0}^{n-1} \frac{(x-a)^{k-q}}{\Gamma(k+1-q)} (f(x)g(x))^{(k)}(a). \qquad (4.2.28)$$

Mittag-Leffler function of order $\alpha > 0$ is defined by the series

$$E_\alpha(x) = \sum_{k=0}^{\infty} \frac{x^k}{\Gamma(\alpha k + 1)}.$$

Mittag-Leffler function was introduced by the Swedish mathematician Mittag-Leffler in 1903, though its pivotal role in the solutions of fractional differential equations is realized only in last two decades. Various generalizations of this function exist in literature, which are quite useful in solving FDEs. Its generalization to two parameters was given by Wiman (1905) which is defined as

$$E_{\alpha,\beta}(x) = \sum_{k=0}^{\infty} \frac{x^k}{\Gamma(\alpha k + \beta)}, \quad \alpha > 0, \ \beta > 0.$$

Solving FDE: Laplace Transform Method

Example 4.2.1. Solve the following initial value problem in the case of the inhomogeneous Bagley–Torvik equation using Laplace transform [5]:

$$D^2 y(x) + D^{\frac{3}{2}} y(x) + y(x) = 1 + x,$$

$$(4.2.29)$$

$$y(0) = y'(0) = 1.$$

Applying Laplace transform to Eq. (4.2.29) we obtain

$$s^2 Y(s) - sy(0) - y'(0) + s^{\frac{3}{2}} Y(s) - s^{\frac{1}{2}} y(0) - s^{\frac{-1}{2}} y'(0) + Y(s)$$

$$= \frac{1}{s} + \frac{1}{s^2},$$

$$Y(s)(s^2 + s^{\frac{3}{2}} + 1) = s + 1 + s^{\frac{1}{2}} + s^{\frac{-1}{2}} + \frac{1}{s} + \frac{1}{s^2},$$

$$Y(s)(s^2 + s^{\frac{3}{2}} + 1) = \frac{s^3 + s^2 + s^{\frac{5}{2}} + s^{\frac{3}{2}} + s + 1}{s^2}$$

$$= (s^2 + s^{\frac{3}{2}} + 1)\left(\frac{1}{s} + \frac{1}{s^2}\right).$$

Therefore

$$Y(s) = \frac{1}{s} + \frac{1}{s^2}. \tag{4.2.30}$$

Taking inverse Laplace transform of (4.2.30) we deduce the exact solution,

$$y(x) = 1 + x.$$

Exercise 4.2.6. Solve the following inhomogeneous linear equation by Laplace transform method:

$$D^\alpha y(x) + y(x) = \frac{2x^{2-\alpha}}{\Gamma(3-\alpha)} - \frac{x^{1-\alpha}}{\Gamma(2-\alpha)} + x^2 - x,$$

$$y(0) = 1, \quad 0 < \alpha \le 1.$$

Exercise 4.2.7. Using term-by-term differ/integration show that

$$I_0^q e^{\lambda t} = t^q E_{1,1+q}(\lambda t), \quad - \tag{4.2.31}$$

$$D_0^q e^{\lambda t} = t^{-q} E_{1,1-q}(\lambda t), \tag{4.2.32}$$

$${}^c D_0^q e^{\lambda t} = \lambda^n t^{n-q} E_{1,1+n-q}(\lambda t). \tag{4.2.33}$$

Exercise 4.2.8. Show by term-by-term fractional-order integration

$$I_0^q [E_{\alpha,\beta}(\lambda t^\alpha) t^{\beta-1}] = t^{\beta+q-1} E_{\alpha,\beta+q}(\lambda t^\alpha), \quad \beta > 0, \quad q > 0. \tag{4.2.34}$$

Exercise 4.2.9. Show that

$$D_0^q [t^{q-1} E_{q,q}(at^q)] = at^{q-1} E_{q,q}(at^q), \tag{4.2.35}$$

$${}^c D_0^q [E_q(at^q)] = aE_q(at^q). \tag{4.2.36}$$

Exercise 4.2.10. Evaluate ${}^c D_0^{2.2}[e^{3t} t^2]$ using Leibniz rule.

Exercise 4.2.11. Show that

$$\mathcal{L}\left[t^{q-1}E_{p,q}(\pm at^p)\right](s) = \frac{s^{p-q}}{s^p \mp a}, \quad \text{for } s > |a|^{1/p}. \quad (4.2.37)$$

Theorem 4.2.5. *The general solution to the following IVP, for* $n - 1 < q < n$

$$D_0^q y(t) - \lambda y(t) = h(t), D_0^{q-k} y(0) = b_k, \quad k = 1, 2, \ldots, n, \quad (4.2.38)$$

is given by the following expression:

$$y(t) = \sum_{k=1}^n b_k t^{q-k} E_{q,q-k+1}(\lambda t^q)$$

$$+ \int_0^t (t-\tau)^{q-1} E_{q,q}[\lambda(t-\tau)^q] h(\tau) d\tau. \quad (4.2.39)$$

The proof follows by taking the Laplace transform and using Eq. (4.2.37).

4.2.4. Existence, uniqueness and stability of solutions

Existence, uniqueness and stability of solutions of nonlinear fractional differential equations involving Caputo derivative has been studied in detail in [6]. Consider the fractional initial value problem (IVP)

$$^c D^\alpha y(t) = f(t, y), y^{(k)}(0) = c_k, \quad k = 1, 2, \ldots, n - 1, \quad (4.2.40)$$

where $n - 1 < \alpha \leq n$. For brevity of notations, we will use $^c D^\alpha y(t)$ and $I^\alpha y(x)$ to denote $^c D_0^\alpha y(t)$ and $I_0^\alpha y(x)$, respectively.

Lemma 4.2.2. *The IVP* (4.2.40) *is equivalent to Volterra integral equation:*

$$y(t) = \sum_{k=0}^{n-1} c_k \frac{t^k}{k!} + I^\alpha f(t, y). \quad (4.2.41)$$

Exercise 4.2.10. Prove Lemma 4.2.2.

Theorem 4.2.6 ([6]). *Let $f : W \to \mathbb{R}^n$ be C^1 where*

$$W = [0, \ \chi^*] \times \prod_{j=1}^{n} [y(0) - l, \ y(0) + l], \quad \chi^* > 0, \ l > 0.$$

Then the IVP (4.2.40) has a unique solution $\bar{y}(t) : [0, \chi] \to \mathbb{R}$, where

$$\chi = \min \left\{ \chi^*, \left(\frac{l\,\Gamma(\alpha+1)}{[1+\alpha]\|f\|} \right)^{1/\alpha}, \left(\frac{l\,k!}{[1+\alpha]|c_k|} \right)^{\frac{1}{k}} \right\}, \quad k = 1, \ldots, n-1,$$

(4.2.42)

and $[1 + \alpha]$ denotes integral part of $1 + \alpha$.

4.2.5. Dependence of solution on initial conditions

Theorem 4.2.7. *Let the functions $f : W \to \mathbb{R}$, where*

$$W = [0, \ \chi^*] \times [y(0) - l, \ y(0) + l], \quad \chi^* > 0, l > 0,$$

be C^1. Let f be Lipschitz in the second variable, i.e.

$$|f(t, \ y(t)) - f(t, \ z(t))| \leq L \, |y(t) - z(t)|. \tag{4.2.43}$$

Let $y(t)$ and $z(t)$ be the solutions of the initial value problems:

$$
\begin{aligned}
{}^c D^\alpha y(t) &= f(t, \ y(t)), \qquad y^{(k)}(0) = c_k, \ 0 \leq k \leq n-1, \\
{}^c D^\alpha z(t) &= f(t, \ z(t)), \qquad z^{(k)}(0) = d_k, \ 0 \leq k \leq n-1, \tag{4.2.44}
\end{aligned}
$$

respectively, where $n - 1 < \alpha \leq n$. Then

$$|y(t) - z(t)| \leq \|T - T'\| \, E_\alpha(Lt^\alpha), \tag{4.2.45}$$

where $T(t) = \sum_{k=0}^{n-1} c_k \frac{t^k}{k!}$ and $T'(t) = \sum_{k=0}^{n-1} d_k \frac{t^k}{k!}$ and E_α is the Mittag-Leffler function.

For similar analysis of FDEs involving Riemann–Liouville derivatives, we refer the reader to [7, 8].

4.3. Decomposition Methods for Solving FDEs

Numerous problems in Physics, Chemistry, Biology and Engineering Science are modeled mathematically by fractional differential equations. Integral transform methods such as Laplace transform, Fourier transform, Mellin transform methods are useful for solving linear FDEs. But for nonlinear FDEs one has to develop other methods. Since most realistic differential equations do not have exact analytic solutions, approximations and numerical techniques, therefore, need to be developed. Recently introduced Adomian Decomposition Method (ADM) [9] has been used extensively for solving a wide range of linear/nonlinear problems. It yields analytical solutions and offers certain advantages over standard numerical methods. It is free from rounding off errors since it does not involve discretization, and is computationally inexpensive. In the present section we describe basics of ADM and illustrate how it is used for solving fractional differential equations. There exists huge literature on this topic. Consider the general functional equation

$$u = f + L(u) + N(u). \tag{4.3.1}$$

Equation (4.3.1) represents a variety of equations such as nonlinear ordinary differential equations, partial differential equations, integral equations, fractional differential equations and systems of these equations. We present here Adomian decomposition method (ADM) introduced by G. Adomian in 1980, which is extensively used in the literature for solving linear/nonlinear FDEs.

4.3.1. Adomian decomposition method

In ADM solution for Eq. (4.3.1) is expressed in the form of infinite series

$$u = \sum_{n=0}^{\infty} u_n, \tag{4.3.2}$$

with $u_0 = f$. Further it is assumed that the nonlinear term $N(u)$ can be expressed as $\sum_{n=0}^{\infty} A_n$ where A_n are called as Adomian polynomials. Thus (4.3.1) becomes

$$\sum_{n=0}^{\infty} u_n = f + L \left(\sum_{n=0}^{\infty} u_n \right) + \sum_{n=0}^{\infty} A_n, \tag{4.3.3}$$

where the Adomian polynomials A_n's are given by the expression

$$A_n = \frac{1}{n!} \left[\frac{\mathrm{d}^k}{\mathrm{d}\lambda^k} N \left(\sum_{k=0}^{n} u_k(x) \lambda^k \right) \right]_{\lambda=0}. \tag{4.3.4}$$

Since L is linear, we can write (4.3.3) as

$$\sum_{n=0}^{\infty} u_n = f + \sum_{n=0}^{\infty} L(u_n) + \sum_{n=0}^{\infty} A_n. \tag{4.3.5}$$

The recursive relation for u_n is thus obtained as follows:

$$u_0 = f$$
$$u_1 = L(u_0) + A_0$$
$$u_2 = L(u_1) + A_1$$
$$\vdots$$
$$u_n = L(u_{n-1}) + A_{n-1}. \tag{4.3.6}$$

The solution to (4.3.1) is always given in the form of k-term approximate solution

$$u = \sum_{n=0}^{k-1} u_n \tag{4.3.7}$$

for suitable integer k.

Exercise 4.3.1. Find Adomian polynomials for the operator $N(u) = u^3$.

Exercise 4.3.2. Find the 4-term approximation for the following IVP using ADM $^cD^{\frac{1}{2}}y = 1 + y^2$, $y(0) = 0$.

Exercise 4.3.3. Write a computer algorithm (Mathematica/ Matlab/Maple) to find Adomian polynomials.

4.3.2. New iterative method

The difficulty in ADM lies in calculating the Adomian's polynomials. As computation of Adomian polynomials in ADM is rather cumbersome, Daftardar-Gejji and Jafari [10] have proposed another decomposition method, called as new iterative method (NIM) which is simple and easy to implement. It is economical in terms of computer power/memory and does not involve tedious calculations such as Adomian polynomials. In many cases it gives analytical solutions and if a closed form solution is not possible, then computation of only first two or three terms gives very good numerical approximation. In this section, we explain NIM and its utility in the context of FDEs.

Let u be a solution of Eq. (4.3.1) having the series form:

$$u = \sum_{i=0}^{\infty} u_i. \qquad (4.3.8)$$

Since L is linear $L\left(\sum_{i=0}^{\infty} u_i\right) = \sum_{i=0}^{\infty} L(u_i)$. The nonlinear operator here is decomposed as:

$$N\left(\sum_{i=0}^{\infty} u_i\right) = N(u_0) + \sum_{i=1}^{\infty}\left\{ N\left(\sum_{j=0}^{i} u_j\right) - N\left(\sum_{j=0}^{i-1} u_j\right)\right\} \qquad (4.3.9)$$

$$= \sum_{i=0}^{\infty} G_i, \qquad (4.3.10)$$

where $G_0 = N(u_0)$ and $G_i = \{N(\sum_{j=0}^{i} u_j) - N(\sum_{j=0}^{i-1} u_j)\}$, $i \geq 1$. Hence Eq. (4.3.1) is equivalent to

$$\sum_{i=0}^{\infty} u_i = f + \sum_{i=0}^{\infty} L(u_i) + \sum_{i=0}^{\infty} G_i. \qquad (4.3.11)$$

Further define the recurrence relation:

$$u_0 = f,$$
$$u_1 = L(u_0) + G_0,$$
$$u_{m+1} = L(u_m) + G_m, \quad m = 1, 2, \ldots. \qquad (4.3.12)$$

Then

$$(u_1 + \cdots + u_{m+1})$$
$$= L(u_0 + \cdots + u_m) + N(u_0 + \cdots + u_m), \quad m = 1, 2, \ldots,$$

and $u = f + \sum_{i=1}^{\infty} u_i$.

For the convergence analysis of this method, we refer the reader to [11]. We present below some illustrative examples to explain the method.

Example 4.3.1. Consider the time-fractional diffusion equation

$$^cD_t^\alpha u(x,t) = u_{xx}(x,t), \quad t > 0, \ x \in \mathbb{R}, \ 0 < \alpha \le 1, \quad (4.3.13)$$
$$u(x,0) = \sin(x). \quad (4.3.14)$$

System (4.3.13)–(4.3.14) is equivalent to

$$u = \sin(x) + I_t^\alpha u_{xx}. \quad (4.3.15)$$

Using the NIM algorithm, we get the recurrence relation

$$u_0 = \sin(x), \quad u_1 = -\sin(x)\frac{t^\alpha}{\Gamma(\alpha+1)}, \cdots$$

In general $u_j = (-1)^j \sin(x)\frac{t^{j\alpha}}{\Gamma(j\alpha+1)}$, $j = 0, 1, 2, \ldots$ The solution of (4.3.13)–(4.3.14) is thus

$$u(x,t) = \sum_{j=0}^{\infty} u_j(x,t) = \sin(x) \sum_{j=0}^{\infty} \frac{(-t^\alpha)^j}{\Gamma(j\alpha+1)}$$
$$= \sin(x)E_\alpha(-t^\alpha).$$

Example 4.3.2. Consider the time-fractional wave equation

$$^cD_t^\alpha u(x,t) = k \cdot u_{xx}(x,t), \quad t > 0, \ x \in \mathbb{R}, \ 1 < \alpha \le 2, \quad (4.3.16)$$
$$u(x,0) = x^2, \ u_t(x,0) = 0. \quad (4.3.17)$$

We get the equivalent integral equation of IVP (4.3.16)–(4.3.17) as

$$u = x^2 + k \cdot I_t^\alpha u_{xx}. \quad (4.3.18)$$

Applying the NIM, we get $u_0 = x^2$, $u_1 = 2k \cdot \frac{t^\alpha}{\Gamma(\alpha+1)}$, $u_2 = 0, \ldots$. The solution of (4.3.16)–(4.3.17) is

$$u(x,t) = \sum_{i=0}^\infty u_i = x^2 + 2k \cdot \frac{t^\alpha}{\Gamma(\alpha+1)}. \qquad (4.3.19)$$

Example 4.3.3. Consider the space-fractional diffusion equation

$$u_t(x,t) = k \cdot {}^c D_x^\beta u(x,t), \quad t > 0, \ x \in \mathbb{R}, \ 1 < \beta \le 2, \qquad (4.3.20)$$

$$u(x,0) = \frac{2x^\beta}{\Gamma(1+\beta)}. \qquad (4.3.21)$$

Integrating (4.3.20) and using (4.3.21) we get

$$u(x,t) = \frac{2x^\beta}{\Gamma(1+\beta)} + k \int_0^t \left(D_x^\beta u(x,t)\right) \mathrm{d}t. \qquad (4.3.22)$$

Applying the NIM, we get

$$u_0 = \frac{2x^\beta}{\Gamma(1+\beta)}, \ u_1 = 2kt, \ u_2 = 0, \ldots.$$

The solution of (4.3.20)–(4.3.21) turns out to be

$$u(x,t) = \frac{2x^\beta}{\Gamma(1+\beta)} + 2kt. \qquad (4.3.23)$$

Exercise 4.3.3. Using NIM show that the IVP ${}^c D_0^q y(t) = \lambda y(t)$, $y^{(k)}(0) = b_k, k = 0, \ldots, n-1$, $n-1 < \alpha \le n$ has the solution $y(t) = \sum_{k=0}^{n-1} b_k t^k E_{q,k+1}(\lambda t^q)$.

Exercise 4.3.4. Find the 3-term approximation given by NIM for the IVP given in Exercise 4.3.2.

For more illustrations to a variety of problems, we refer the reader to [3].

4.3.3. Numerical methods for solving FDEs

Fractional Adams–Bashforth–Moulton method

Consider the initial value problem (IVP) for $0 < \alpha < 1$:

$${}^c D_0^\alpha x(t) = f(t, x(t)), \quad x(0) = x_0, \qquad (4.3.24)$$

where $^cD_0^\alpha$ denotes Caputo derivative and $f : [0, T] \times \mathbb{D} \longrightarrow \mathbb{R}$, $\mathbb{D} \subseteq \mathbb{R}$. For solving Eq. (4.3.24) on $[0, T]$, the interval is divided into l subintervals. Consider an equi-spaced grid with step length h; $t_j = jh, j = 0, 1, \ldots$. Let x_j denote the approximate solution at t_j and $x(t_j)$ denote the exact solution of the IVP (4.3.24) at t_j.

Fractional rectangular formula

The solution of the IVP (4.3.24) at the point t_n is

$$x(t_n) = x_0 + \frac{1}{\Gamma(\alpha)} \int_0^{t_n} (t_n - \tau)^{\alpha-1} f(\tau, x(\tau)) d\tau. \qquad (4.3.25)$$

On each subinterval $[t_k, t_{k+1}]$, $k = 0, \ldots, n-1$, the function $f(t, x)$ is approximated by constant value $f(t_k, x(t_k))$ to obtain

$$x(t_n) = x_0 + \frac{1}{\Gamma(\alpha)} \sum_{k=0}^{n-1} \int_{t_k}^{t_{k+1}} (t_n - s)^{\alpha-1} f(s, x(s)) ds$$

$$\approx x_0 + \frac{1}{\Gamma(\alpha)} \sum_{k=0}^{n-1} \int_{t_k}^{t_{k+1}} (t_n - s)^{\alpha-1} f_k ds,$$

where $f_k = f(t_k, x_k)$. Hence fractional rectangular formula yields

$$x_n = x_0 + h^\alpha \sum_{k=0}^{n-1} b_{n-k-1} f_k, \quad \text{where } b_k = \frac{[(k+1)^\alpha - k^\alpha]}{\Gamma(\alpha+1)}. \qquad (4.3.26)$$

Fractional trapezoidal formula

In this method on each subinterval $[t_k, t_{k+1}]$, the function $f(t)$ is approximated by straight line as

$$\tilde{f}(t, x(t)) \mid_{[t_k, t_{k+1}]} = \frac{t_{k+1} - t}{t_{k+1} - t_k} f(t_k, x(t_k))$$

$$+ \frac{t - t_k}{t_{k+1} - t_k} f(t_{k+1}, x(t_{k+1})).$$

$$I_0^\alpha f(t_n, x(t_n)) \approx \frac{1}{\Gamma(\alpha)} \sum_{k=0}^{n-1} \int_{t_k}^{t_{k+1}} (t_n - t)^{(\alpha-1)} \tilde{f}(t, x(t)) \mid_{[t_k, t_{k+1}]} dt$$

$$= h^\alpha \sum_{k=0}^{n} a_{n-k} f(t_k, x(t_k)),$$

where

$$a_j = \begin{cases} \dfrac{1}{\Gamma(\alpha+2)} & \text{if } j = 0, \\[2ex] \dfrac{(j-1)^{\alpha+1} - 2j^{\alpha+1} + (j+1)^{\alpha+1}}{\Gamma(\alpha+2)} & \text{if } j = 1, \ldots, n-1, \\[2ex] \dfrac{(n-1)^{\alpha+1} - n^\alpha(n - \alpha - 1)}{\Gamma(\alpha+2)} & \text{if } j = n. \end{cases}$$

Hence fractional trapezoidal formula yields

$$x_n = x_0 + h^\alpha \sum_{j=0}^{n} a_{n-j} f(t_j, x_j), \tag{4.3.27}$$

where a_j's are as defined above.

Fractional rectangle formula and fractional trapezoidal formula form a predictor–corrector algorithm. A preliminary approximation x_n^p (predictor) is made using Eq. (4.3.26), which is substituted in Eq. (4.3.27) to give a corrector.

Thus the fractional Adams–Bashforth–Moulton formula is

$$x_n^p = x_0 + h^\alpha \sum_{j=0}^{n-1} b_{n-j-1} f(t_j, x_j), \tag{4.3.28}$$

$$x_n^c = x_0 + h^\alpha \sum_{j=0}^{n-1} a_{n-j} f(t_j, x_j) + h^\alpha a_0 f(t_n, x_n^p). \tag{4.3.29}$$

New predictor–corrector method

Daftardar-Gejji *et al.* [15, 16] have developed a new predictor–corrector formula which is derived by combining fractional trapezoidal formula and NIM [6] for solving the IVP (4.3.24) numerically.

Fractional trapezoidal formula given by Eq. (4.3.27) can be written as

$$x(t_n) = x(0) + h^\alpha \sum_{j=0}^{n} a_{n-j} f(t_j, x_j)$$

$$= x(0) + h^\alpha \sum_{j=0}^{n-1} a_{n-j} f(t_j, x_j) + \frac{h^\alpha}{\Gamma(\alpha+2)} f(t_n, x_n). \quad (4.3.30)$$

The solution of Eq. (4.3.30) can be approximated by NIM where

$$N(x(t_n)) = \frac{h^\alpha}{\Gamma(\alpha+2)} f(t_n, x_n). \quad (4.3.31)$$

We apply NIM to get approximate value of x_1 as follows:

$$x(t_1) = x_1 = x_0 + h^\alpha a_1 f(t_0, x_0) + \frac{h^\alpha}{\Gamma(\alpha+2)} f(t_1, x_1), \quad (4.3.32)$$

$$x_{1,0} = x_0 + h^\alpha a_1 f(t_0, x_0),$$

$$x_{1,1} = N(x_{1,0}) = \frac{h^\alpha}{\Gamma(\alpha+2)} f(t_1, x_{1,0}),$$

$$x_{1,2} = N(x_{1,0} + x_{1,1}) - N(x_{1,0}).$$

The 3-term approximation is given by $x_1 \approx x_{1,0} + x_{1,1} + x_{1,2} = x_{1,0} + N(x_{1,0} + x_{1,1})$. This gives a new predictor–corrector formula as follows:

$$y_1^p = x_{1,0}, \quad z_1^p = N(x_{1,0})$$

$$x_1^c = y_1^p + \frac{h^\alpha}{\Gamma(\alpha+2)} f(t_1, y_1^p + z_1^p),$$

$x(t_2), x(t_3), \ldots$ can be obtained similarly.

3-Term new predictor–corrector formula

The 3-term new predictor–corrector method (3-term NPCM) consists of the following formulas:

$$y_n^p = x_0 + h^\alpha \sum_{j=0}^{n-1} a_{n-j} f(t_j, x_j),$$

$$z_n^p = \frac{h^\alpha}{\Gamma(\alpha+2)} f(t_n, y_n^p),$$

$$x_n^c = y_n^p + \frac{h^\alpha}{\Gamma(\alpha+2)} f(t_n, y_n^p + z_n^p).$$

Here y_n^p and z_n^p are called as predictors and x_n^c is the corrector.

For convergence and stability analysis of this method, we refer the reader to [15, 16]. Numerous examples have been solved by this method and the CPU time required for calculations is compared with other methods, from which it can be concluded that this method is more accurate and time efficient compared to existing methods in the literature [15]. It requires only half the time taken by fractional Adams–Bashforth–Moulton method, as only a_j's are involved and computations related to b_j's are not needed. Further it has been proved that the 3-term NPCM has better stability properties as compared to fractional Adams–Bashforth–Moulton method [15].

Further this method is extended to solve fractional differential equations involving delay [16].

Exercise 4.3.5. Consider the following fractional-order differential equation [15]

$$^cD_0^\alpha y(x) + y^4(x) = \frac{\Gamma(2\alpha+1)x^\alpha}{\Gamma(\alpha+1)} - \frac{2x^{2-\alpha}}{\Gamma(3-\alpha)} + (x^{2\alpha}-x^2)^4; \quad y(0) = 0.$$

(4.3.33)

Exact solution of the IVP (4.3.33) is $x^{2\alpha} - x^2$. Solve this example by both the methods described above and compare the solutions obtained with the exact solution for accuracy. Also compare the CPU time taken for the simulations.

4.4. Fractional Boundary Value Problems

The time fractional diffusion-wave equation is obtained from the classical diffusion or wave equation by replacing the first- or second-order time derivative by a fractional derivative of order with $0 < \alpha < 1$ or $1 < \alpha < 2$, respectively. It represents anomalous subdiffusion if $0 < \alpha < 1$, and anomalous superdiffusion in case of $1 < \alpha < 2$. It is a well-established fact that this equation models various phenomena.

Nigmatullin [17] has employed the fractional diffusion equation to describe diffusion in media with fractal geometry. Mainardi [18] has shown that the fractional wave equation governs the propagation of mechanical diffusive waves in viscoelastic media. Metzler and Klafter [19] have demonstrated that fractional diffusion equation describes a non-Markovian diffusion process with a memory. Daftardar-Gejji *et al.* [20, 21] have used the method of separation of variables to solve fractional BVP. In this section we illustrate how the method of separation of variables can be extended to solve fractional diffusion-wave equation under homogeneous/non-homogeneous boundary conditions to get analytical solutions.

Caputo (partial) fractional derivative with respect to t is defined as

$$D_t^\alpha u(x,t) = \frac{1}{\Gamma(m-\alpha)} \int_0^t (t-\tau)^{m-\alpha-1} \frac{\partial^m u(x,t)}{\partial t^m} d\tau,$$

$$m-1 < \alpha \le m, m \in \mathbb{N}, t > 0. \quad (4.4.1)$$

Consider the following non-homogeneous fractional diffusion-wave equation:

$$D_t^\alpha u(x,t) = k\frac{\partial^2 u(x,t)}{\partial x^2}, \quad 0 < x < \pi, t > 0, 0 < \alpha \le 1, \quad (4.4.2)$$

$$u(0,t) = u(\pi,t) = 0, \quad t \ge 0, \quad (4.4.3)$$

$$u(x,0) = f(x), \quad 0 < x < \pi. \quad (4.4.4)$$

We explain how the method of separation of variables can be used to solve the fractional BVP. Assume $u(x,t) = X(x)T(t)$, then (4.4.2), along with condition (4.4.3), yields

$$X''(x) + \lambda X(x) = 0, \quad X(0) = X(\pi) = 0, \quad (4.4.5)$$

and

$$D_t^\alpha T(t) + \lambda k T(t) = 0, \quad t \ge 0. \quad (4.4.6)$$

The Stürm–Liouville problem given by (4.4.5) has eigenvalues $\lambda_n = n^2$ and the corresponding eigenfunctions $X_n(x) = \sin nx$ ($n = 1, 2, \ldots$). The solution of (4.4.6) for the case $\lambda = n^2$ is (up to a constant multiple) $T_n(t) = E_\alpha(-n^2kt^\alpha)$, where E_α denotes the Mittag-Leffler function. Moreover

$$u(x,t) = \sum_{n=1}^{\infty} C_n X_n(x) T_n(t),$$

where C_n's are determined by the condition (4.4.4),

$$\sum_{n=1}^{\infty} C_n \sin nx = f(x), \quad 0 < x < \pi,$$

which yields

$$C_n = \frac{2}{\pi} \int_0^\pi f(r) \sin nr \, dr.$$

Hence

$$u(x,t) = \frac{2}{\pi} \sum_{n=1}^{\infty} E_\alpha(-n^2kt^\alpha) \sin nx \int_0^\pi f(r) \sin nr \, dr.$$

Exercise 4.4.1. Extend the method of separation of variables to non-homogeneous BVP and solve

$$D_t^\alpha u(x,t) = k \frac{\partial^2 u(x,t)}{\partial x^2} + q(t), \quad 0 < x < \pi, \, t > 0, \, 0 < \alpha \le 2,$$
$$(4.4.7)$$

under the BC given by the Eqs. (4.4.3)–(4.4.4).

This method has been extended to solve multi-order fractional BVPs as well [21].

4.5. Systems of FDEs

4.5.1. Linear case

Mittag-Leffler function for matrix arguments

A generalization of Mittag-Leffler function for matrix arguments can be defined as

$$E_\alpha(A) = \sum_{k=0}^{\infty} \frac{A^k}{\Gamma(\alpha k + 1)}, \quad (4.5.1)$$

where A is an $n \times n$ real matrix. It is easy to show that this series converges absolutely for all square matrices in the uniform norm, where uniform norm of $n \times n$ matrix A is defined to be

$$\|A\| = \max \left\{ \frac{|A(x)|}{|x|} \leq 1 \right\}. \tag{4.5.2}$$

Analysis of fractional-order differential equations involving Caputo derivatives has been studied by Daftardar-Gejji and Jafari [5]. We present here some important results.

Theorem 4.5.1. *The unique solution to the system*

$$^cD_0^\alpha \bar{y}(t) = A\,\bar{y}(t), \quad \bar{y}(0) = \bar{c},$$
$$0 < \alpha \leq 1, \quad \bar{c} \in \mathbb{R}^n \tag{4.5.3}$$

is $E_\alpha(t^\alpha A)\bar{c}$, where A is an $n \times n$ real matrix and $\bar{y} = [y_1, \ldots, y_n]^t$.

Proof. We express left-hand-side of (4.5.3)

$$^cD_0^\alpha \bar{y}(t) = D^\alpha [\bar{y}(t) - \bar{c}] \tag{4.5.4}$$

$$= D^\alpha [E_\alpha (t^\alpha A)\,\bar{c} - \bar{c}] \tag{4.5.5}$$

$$= D^\alpha \left[\left\{ I + \frac{t^\alpha A}{\Gamma(\alpha + 1)} + \frac{(t^\alpha A)^2}{\Gamma(2\alpha + 1)} + \cdots \right\} \bar{c} - I\bar{c} \right] \tag{4.5.6}$$

$$= D^\alpha \left[\left\{ \frac{t^\alpha A}{\Gamma(\alpha + 1)} + \frac{(t^\alpha A)^2}{\Gamma(2\alpha + 1)} + \cdots \right\} \bar{c} \right]. \tag{4.5.7}$$

But the series

$$\left\{ \frac{t^\alpha A}{\Gamma(\alpha + 1)} + \frac{(t^\alpha A)^2}{\Gamma(2\alpha + 1)} + \cdots \right\}$$

is uniformly convergent on $[0, \chi]$ as

$$\left\| \frac{(t^\alpha A)^k}{\Gamma(k\,\alpha + 1)} \right\| \leq \frac{\|(\chi^\alpha A)^k\|}{\Gamma(k\,\alpha + 1)}$$

and the series $\sum_{k=1}^{\infty} \frac{\|(\chi^\alpha A)^k\|}{\Gamma(k\alpha+1)}$ is convergent. Hence

$$
{}^c D_0^\alpha \bar{y}(t) = D^\alpha \left(\sum_{n=1}^{\infty} \frac{(t^\alpha A)^n}{\Gamma(1+\alpha n)} \right) \bar{c} = \left(\sum_{n=1}^{\infty} \frac{(t^\alpha)^{n-1} A^n}{\Gamma(1+\alpha(n-1))} \right) \bar{c}
$$

$$
= \left(\sum_{n=0}^{\infty} \frac{(t^\alpha)^n A^{n+1}}{\Gamma(1+\alpha n)} \right) \bar{c} = A \left(\sum_{n=0}^{\infty} \frac{(t^\alpha A)^n}{\Gamma(1+\alpha n)} \right) \bar{c}
$$

$$
= A\, E_\alpha(t^\alpha A)\bar{c} = A\, \bar{y}(t). \tag{4.5.8}
$$

Lemma 4.5.1. *If V is an eigenvector of A with eigenvalue λ, then V is also an eigenvector of $E_\alpha(A)$ with eigenvalue $E_\alpha(\lambda)$.*

Exercise 4.5.1. Prove Lemma 4.5.1.

Theorem 4.5.2. *If V is an eigenvector of the matrix A corresponding to the eigenvalue λ, then $X(t) = E_\alpha(\lambda t^\alpha)V$ is a solution of the equation ${}^c D_0^\alpha \bar{y}(t) = A\, \bar{y}(t)$.*

Exercise 4.5.2. Prove Theorem 4.5.2.

Exercise 4.5.3. Solve the system of equations

$$
{}^c D^\alpha y_1 = y_1 + 3y_2,
$$
$$
{}^c D^\alpha y_2 = y_1 - y_2. \tag{4.5.9}
$$

Hint: In view of Theorem 4.5.2, $X_1(t) = E_\alpha(2t^\alpha)\binom{3}{1}$, $X_2(t) = E_\alpha(-2t^\alpha)\binom{1}{-1}$ are two linearly independent solutions. A general solution is a linear combination of these two solutions and the initial conditions will determine the arbitrary constants.

Exercise 4.5.4. Solve the following IVP for the initial conditions: $y_1(0) = 1$, $y_2(0) = 0$.

$$
{}^c D^\alpha y_1 = y_1 + 2y_2,
$$
$$
{}^c D^\alpha y_2 = 4y_1 - y_2. \tag{4.5.10}
$$

Exercise 4.5.5. Solve the system $^cD^\alpha x = Ax, x(0) = [1, 2, 0]^t$ where

$$A = \begin{bmatrix} \lambda & 1 & 0 \\ 0 & \lambda & 1 \\ 0 & 0 & \lambda \end{bmatrix}.$$

Stability analysis

We now turn our attention to the question of stability analysis of linear system of FDEs. We state below a basic result due to Matignon [22, 23].

Theorem 4.5.3. *Consider the system of fractional differential equations $^cD_0^\alpha \bar{y}(t) = A\,\bar{y}(t)$, where A is an arbitrary $n \times n$ matrix.*

(1) *The solution $y = 0$ of the system is asymptotically stable if and only if all the eigenvalues λ_j of A satisfy $|\arg \lambda_j| > \alpha \pi/2$.*

(2) *The solution $y = 0$ is stable if and only if the eigenvalues satisfy $|\arg \lambda_j| \geq \alpha \pi/2$ and all eigenvalues with $|\lambda_j| = \alpha \pi/2$ have a geometric multiplicity that coincides with their algebraic multiplicity.*

Thus, there exists a threshold value α^*, say, such that the system is asymptotically stable if $\alpha < \alpha^*$ and unstable if $\alpha > \alpha^*$. In other words, the stability properties can be improved by reducing the order α of the differential operator. An analogous statement applies to nonlinear fractional differential equations where systems tend to exhibit chaotic behavior if the order of the differential operators is larger than the threshold value α^* and remain stable if the order is less than α^*. We will revisit this concept in the follwing section which deals with chaos in nonlinear fractional ordered systems.

4.5.2. Chaos in nonlinear dynamics

Chaos is aperiodic behavior in a deterministic system which exhibits high sensitivity to initial conditions. Hence the deterministic nature of these systems does not make them predictable. This behavior is known as deterministic chaos. Such behavior was first observed by Edward Lorenz (1961) while solving a system of three equations governing weather prediction on a computer. He observed that small changes in initial conditions produce large changes in the long term outcome of the following system of equations, popularly known as Lorenz equations

$$\dot{x} = \sigma\left(y - x\right),$$
$$\dot{y} = rx - y - xz, \qquad\qquad (4.5.11)$$
$$\dot{z} = xy - \mu z,$$

where σ, r, μ are constants. Lorenz observed that the system behaves chaotically whenever $\sigma = 10$, $\mu = 8/3$ and $r > 24.74$. A chaotic system has solutions that remain bounded but never converge to a fixed point or a periodic orbit.

4.5.3. Phase portraits (Lorenz system)

Figure 4.1 shows time-series $x(t)$, $y(t)$ and $z(t)$ and Figs. 4.2–4.4 show different phase portraits for the case $\sigma = 10$, $\mu = 8/3$ and $r = 28$.

4.5.4. Fractional-ordered systems and chaos

The development of a qualitative theory of FDE is still in its early infancy. One of the reasons for this is that fractional differential equations do not generate semi-groups and the existing qualitative

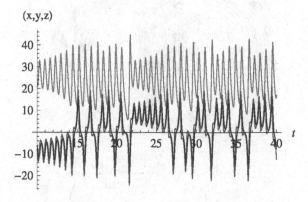

Figure 4.1.   Time series-Lorenz system.

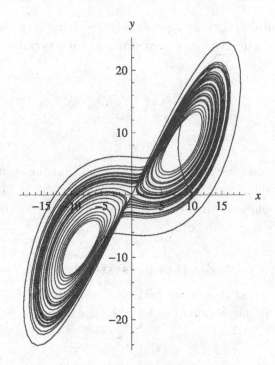

Figure 4.2. *xy*-Phase portrait.

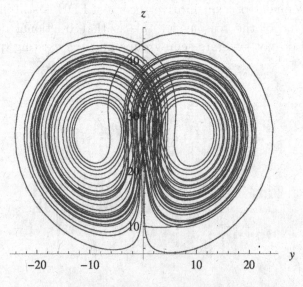

Figure 4.3. *yz*-Phase portrait.

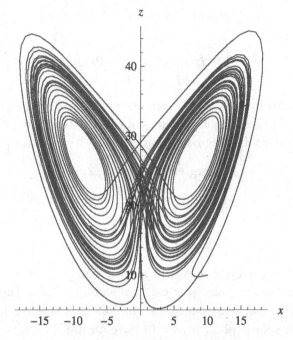

Figure 4.4. xz-Phase portrait.

theory of dynamical systems cannot be extended directly to the fractional case.

Consider the fractional-ordered dynamical system

$$D^\alpha x_i = f_i(x_1, x_2, x_3), \quad 1 \le i \le 3, \qquad (4.5.12)$$

where $0 < \alpha < 1$. $p = (x_1^*, x_2^*, x_3^*)$ is called an equilibrium point of system (4.5.12) if $f_i(p) = 0$ for $i = 1, 2, 3$. Let $p \equiv (x_1^*, x_2^*, x_3^*)$ be an equilibrium point of the system (4.5.12) and $\xi_i = x_i - x_i^*$, $1 \le i \le 3$, a small perturbation from a fixed point:

$$D^\alpha \xi_i \approx \xi_1 \frac{\partial f_i(p)}{\partial x_1} + \xi_2 \frac{\partial f_i(p)}{\partial x_2} + \xi_3 \frac{\partial f_i(p)}{\partial x_3}, \quad 1 \le i \le 3. \qquad (4.5.13)$$

System (4.5.13) is equivalent to

$$\begin{pmatrix} D^\alpha \xi_1 \\ D^\alpha \xi_2 \\ D^\alpha \xi_3 \end{pmatrix} = J \begin{pmatrix} \xi_1 \\ \xi_2 \\ \xi_3 \end{pmatrix}, \qquad (4.5.14)$$

where J is the Jacobian matrix evaluated at point p.

$$J = \begin{pmatrix} \partial_1 f_1(p) & \partial_2 f_1(p) & \partial_3 f_1(p) \\ \partial_1 f_2(p) & \partial_2 f_2(p) & \partial_3 f_2(p) \\ \partial_1 f_3(p) & \partial_2 f_3(p) & \partial_3 f_3(p) \end{pmatrix}. \tag{4.5.15}$$

An equilibrium point p of the system (4.5.12) is locally asymptotically stable if all the eigenvalues of the Jacobian matrix evaluated at p satisfy the following condition:

$$|\arg(\mathrm{eig}(J))| > \alpha\pi/2. \tag{4.5.16}$$

An equilibrium point p is defined as a non-hyperbolic equilibrium point if

$$|\arg(\mathrm{eig}(J))| \neq \alpha\pi/2,$$

for every eigenvalue the Jacobian matrix J at p.

An equilibrium point p is called a saddle point if the Jacobian matrix at p has at least one stable and one unstable eigenvalue (see Fig. 4.5). A saddle point is said to have an index one (two) if there is exactly one (two) unstable eigenvalue(s). A necessary condition

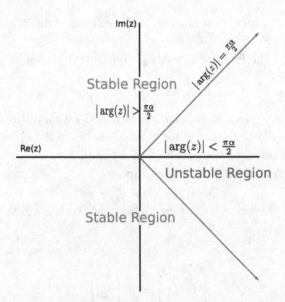

Figure 4.5. Stable and unstable regions.

for the fractional-order system (4.5.12) to remain chaotic is that at least one of the eigenvalues do not satisfy the condition (4.5.16). It is established in the literature that, scrolls are generated only around the saddle points of index two. Saddle points of index one are responsible only for connecting scrolls.

In a seminal paper [24], Grigorenko and Grigorenko demonstrated existence of chaotic solutions in fractional ordered Lorenz dynamical system. Consider the fractional-order Lorenz system [24, 25]

$$D^{\alpha_1} x = 10(y - x),$$
$$D^{\alpha_2} y = 28x - y - xz, \qquad\qquad (4.5.17)$$
$$D^{\alpha_3} z = xy - \frac{8}{3} z,$$

where $0 < \alpha_i \leq 1\,(i = 1, 2, 3)$. If $\alpha_1 = \alpha_2 = \alpha_3$ then we call the system (4.5.17) as commensurate system otherwise incommensurate. Define $\Sigma = \alpha_1 + \alpha_2 + \alpha_3$ as a system order. It is shown [24, 25] using numerical experiments that the chaos exist for the system order $\Sigma < 3$. System is chaotic for the commensurate order $\alpha_1 = \alpha_2 = \alpha_3 = 0.99$ and stable for $\alpha_1 = \alpha_2 = \alpha_3 = 0.98$ (see Fig. 4.6).

One of the important question being addressed herewith is: what is the minimum effective dimension in a fractional-order dynamical system for which the system remains chaotic? The effective dimension being defined as the sum of orders of all involved derivatives. The minimum effective dimension has been numerically calculated for various systems including fractional-order Lorenz system, fractional-order Chua system, fractional-order Rossler system, fractional-order Newton–Leipnik system and so on. Daftardar-Gejji and Bhalekar [26] have studied fractional-order Liu system and have found the minimum effective dimension for the commensurate order 2.76 and incommensurate order system 2.60.

The study of fractional chaotic systems (FCS) has applications in secure communications. Secure codes can be made using FCS which are difficult to break. Fractional-order derivative acts as additional parameter which works as a key. For more details and latest update on this topic the reader may refer to [27].

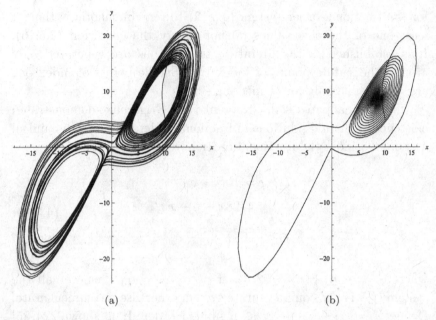

Figure 4.6.   (a) $\alpha_1 = \alpha_2 = \alpha_3 = 0.99$; (b) $\alpha_1 = \alpha_2 = \alpha_3 = 0.98$.

Though much numerical work has been carried out to understand fractional-order dynamical systems, analytical results obtained are very few. The main obstacle in generalizing the results to fractional-ordered systems is that the solutions fail to satisfy semi-group property and hence do not generate *flow* in the traditional sense. In spite of this limitation some important analytical results have been proved such as linearization theorem for fractional systems [28], local stable manifold theorem for planar fractional differential systems [30], non-existence of periodic solutions in case of FDEs [29]. In pursuance to this topic, Deshpande and Daftardar-Gejji [31, 32] have recently proved existence of local stable manifold around an hyperbolic equilibrium point of a fractional system which is summarized below.

Consider the fractional-order dynamical system ($0 \le \alpha \le 1$);

$$D^\alpha x(t) = Ax(t) + f(x), \quad x(0) = x_0 \in \mathbb{R}^n, A \in \mathbb{R}^{n \times n}, \quad (4.5.18)$$

where $f \in C^1[\mathbb{R}^n, \mathbb{R}^n]$, $Df(0) = 0$. Then we have the following lemma.

Lemma 4.1. *IVP* (4.5.18) *is equivalent to the following integral equation:*

$$\phi_t(x_0) := x(t)$$

$$= E_\alpha(t^\alpha A)x_0 + \int_0^t (t - \tau)^{\alpha-1} E_{\alpha,\alpha}(A(t - \tau)^\alpha) f(x(\tau)) d\tau.$$

$$(4.5.19)$$

Let origin be a hyperbolic equilibrium point of system (4.5.18). Then we have the following definition.

Definition 4.1. Local stable set of neighborhood $N_r(0)$, $r > 0$, is defined as

$$W_{loc}^s(N_r(0)) := \left\{ x_0 \in N_r(0) : \phi_t(x_0) \in N_r(0), \forall t \geq 0, \right.$$

$$\left. \text{and} \lim_{t \to \infty} \|\phi_t(x_0)\| = 0 \right\}. \quad (4.5.20)$$

Definition 4.2.

$$E^s = \{ x_0 \in \mathbb{R}^n : \|E_\alpha(t^\alpha A)x_0\| < \epsilon \text{ for some } \epsilon > 0, \forall t \geq 0 \}.$$

Theorem 4.1 (Stable Manifold Theorem). *Let* $f \in C[U, \mathbb{R}^n]$ *where U is a neighborhood of origin, $f(0) = 0$, $Df(0) = 0$. For the fractional system*

$$D^\alpha x(t) = Ax(t) + f(x(t)),$$

$$x(0) = x_0 \in \mathbb{R}^n, \ A \in \mathbb{R}^{n \times n}, \ 0 < \alpha < 1, \quad (4.5.21)$$

with origin as an hyperbolic equilibrium point, there exists $r > 0$ such that for $N_r(0)$, origin belongs to $W_{loc}^s(N_r(0))$ and $W_{loc}^s(N_r(0))$ forms a Lipschitz graph over E^s.

Bibliography

[1] K. B. Oldham and J. Spanier, *The Fractional Calculus: Theory and Applications of Differentiation and Integration to Arbitrary Order*, Dover Publications, 1974.

[2] I. Podlubny, *Fractional Differential Equations*, Academic Press, San Diego, 1999.

[3] V. Daftardar-Gejji (ed.), *Fractional Calculus: Theory and Applications*, Narosa, 2014.

[4] M. Caputo, Linear models of dissipation whose Q is almost frequency independent II, *Geophys. J. Roy. Astron. Soc.* **13** (1967), 529–539.

[5] S. Kazem, Exact solutions of some linear fractional differential equations by Laplace transform, *Int. J. Nonlinear Sci.* **16** (2013), 3–11.

[6] V. Daftardar-Gejji and H. Jafari, Analysis of a system of non autonomous fractional differential equations involving Caputo derivatives, *J. Math. Anal. Appl.*, **328**(2) (2007), 1026–1033.

[7] K. Diethelm and N.J. Ford, Analysis of fractional differential equations, *J. Math. Anal. Appl.* **265** (2002), 229–248.

[8] V. Daftardar-Gejji and A. Babakhani, Analysis of a system of fractional differential equations, *J. Math. Anal. Appl.* **293**(2) (2004), 511–522.

[9] G. Adomian, *Solving Frontier Problems of Physics: The Decomposition Method*, Kluwer Academic Publishers, 1994.

[10] V. Daftardar-Gejji and H. Jafari, An iterative method for solving nonlinear functional equations, *J. Math. Anal. Appl.* **316**(2)(2006), 753–763.

[11] S. Bhalekar and V. Daftardar-Gejji, Convergence of the new iterative method, *Int. J. Differential Equations* (2011); doi:10.1155/2011/989065.

[12] V. Daftardar-Gejji and H. Jafari, Adomian decomposition: a tool for solving a system of fractional differential equations, *J. Math. Anal. Appl.* **301** (2005), 508–518.

[13] K. Diethelm, N. J. Ford and A. D. Freed, A predictor–corrector approach for the numerical solution of fractional differential equations, *Nonlinear Dynamics* **29** (2002), 3–22.

[14] K. Diethelm, N. J. Ford and A. D. Freed, Detailed error analysis for a fractional Adams method, *Numer. Algorithms* **36** (2004), 31–52.

[15] V. Daftardar-Gejji, Y. Sukale and S. Bhalekar, A new predictorcorrector method for fractional differential equations, *Appl. Math. Comput.* **244** (2014), 158–182.

[16] V. Daftardar-Gejji, Y. Sukale and S. Bhalekar Solving fractional delay differential equations: a new approach, *Fract. Calc. Appl. Anal.* **18**(2) (2015), 400–418.

[17] R. R. Nigmatullin, Realization of the generalized transfer equation in a medium with fractal geometry, *Physica B* **133** (1986), 425–430.

[18] F. Mainardi, Fractional diffusive waves in viscoelastic solids, in: *Nonlinear Waves in Solids*, eds. J.L. Wegner and F.R. Norwood, pp. 93–97, Fairfield, 1995.

[19] R. Metzler and J. Klafter, Boundary value problems for fractional diffusion equations, *Physica A* **278** (2000), 107–125.

[20] V. Daftardar-Gejji and H. Jafari, Boundary value problems for fractional diffusion-wave equation, *Australian J. Math. Anal. Appl.* **3** (2006).

[21] V. Daftardar-Gejji and S. Bhalekar, Boundary value problems for multi-term fractional differential equations, *J. Math. Anal. Appl.* **345** (2008), 754–765.

[22] D. Matignon, Stability results for fractional differential equations with applications to control processing, in: *Computational Engineering in Systems and*

Application Multiconference, Vol. 2, *IMACS, IEEE-SMC Proceedings*, Lille, France, July 1996, pp. 963–968.

[23] K. Diethelm, *The Analysis of Fractional Differential Equations*, Springer, 2004.

[24] I. Grigorenko and E. Grigorenko, Chaotic dynamics of the fractional Lorenz system, *Phys. Rev. Lett.* **91**(3) (2003), 034101.

[25] X. Wu and S. Shen, Chaos in the fractional-order Lorenz system, *Int. J. Comput. Math.* **86**(7) (2009), 1274–1282.

[26] V. Daftardar-Gejji and S. Bhalekar, Chaos in fractional ordered Liu system, *Comput. Math. Appl.* **59**(3) (2010), 1117–1127.

[27] R. Caponetto, G. Dongola, L. Fortuna and I. Petras, *Fractional Order Systems: Modeling and Control Applications*, World Scientific, 2010.

[28] L. Changpin and M. Yutian, *Nonlinear Dynamics* **71**(4) (2013), 621–633.

[29] E. Kaslik and S. Sivasundaram, Nonlinear dynamics and chaos in fractional-order neural networks, *Neural Netw.* **32** (2012), 245–256.

[30] N. D. Cong, T. S. Doan, S. Siegmund and H.T. Tuan, On stable manifolds for planar fractional differential equations, *Appl. Math. Comput.* **226** (2014), 157–168.

[31] A. Deshpande and V. Daftardar-Gejji, Local stable manifold theorem for fractional systems, *Nonlinear Dynamics* **83**(4) (2016), 2435–2452.

[32] A. Deshpande and V. Daftardar-Gejji, Erratum to: Local stable manifold theorem for fractional systems, *Nonlinear Dynamics* (2017); http://dx.doi.org/10.1007/s11071-017-3352-1.

Chapter 5

Kober Fractional Calculus and Matrix-Variate Functions*

5.1. Introduction

Matrix-variate functions are real-valued scalar functions of matrix argument, the argument matrix could be real or in the complex domain. Starting from 2009, Mathai [2, 3] has extended fractional calculus from real scalar functions of real scalar variables to matrix-variate functions where the argument is a real positive definite matrix. For real scalar variable case, the basic Riemann–Liouville fractional integrals are the following: The left-sided or first kind fractional integral of order α is given by

$$D_{1,(a,x)}^{-\alpha} f = \frac{1}{\Gamma(\alpha)} \int_a^x (x - t)^{\alpha - 1} f(t) \mathrm{d}t, \quad \Re(\alpha) > 0 \qquad (5.1.1)$$

and the right-sided or second kind fractional integral of order α is given by

$$D_{2,(x,b)}^{-\alpha} f = \frac{1}{\Gamma(\alpha)} \int_x^b (t - x)^{\alpha - 1} f(t) \mathrm{d}t, \quad \Re(\alpha) > 0, \qquad (5.1.2)$$

where $f(t)$ is an arbitrary function. When $a = -\infty$ and $b = \infty$, one has the corresponding Weyl fractional integrals of the first and second kind of order α.

---

*This chapter is summarized from the lectures given by Professor Dr A.M. Mathai.

Let T and X be $p \times p$ real positive definite matrices and let $\Gamma_p(\alpha)$ be the real matrix-variate gamma defined in Chapter 2, Eq. (2.4.3), for $\Re(\alpha) > \frac{p-1}{2}$. Then the extensions of (5.1.1) and (5.1.2) are the following: This author had called the matrix-variate extensions also Riemann–Liouville fractional integrals for the matrix-variate cases in the real and complex domains. The left-sided or first kind Riemann–Liouville fractional integral of order α in the real positive definite $p \times p$ matrix-variate case is defined as

$$D_{1,(A,X)}^{-\alpha} f = \frac{1}{\Gamma_p(\alpha)} \int_{A<T<X} |X - T|^{\alpha - \frac{p+1}{2}} f(T) \mathrm{d}T, \quad \Re(T) > \frac{p-1}{2},$$

(5.1.3)

where $T < X$ means $X - T > O$ (positive definite), $T > O, A > O, X > O, T - A > O$. Here $|(\cdot)|$ means the determinant of $(\cdot)$. $f(T)$ is a real-valued scalar function of the $p \times p$ real positive definite matrix T. The integration is for all T such that $X - T > O$ where X is a fixed $p \times p$ positive definite matrix. The right-sided or second kind Riemann–Liouville fractional integral of order α is defined as

$$D_{2,(X,B)}^{-\alpha} f = \frac{1}{\Gamma_p(\alpha)} \int_{T>X} |T - X|^{\alpha - \frac{p+1}{2}} f(T) \mathrm{d}T, \quad \Re(\alpha) > \frac{p-1}{2},$$

(5.1.4)

where the integral is over $T > O, X > O, T - X > O$. If T is bounded above by a constant positive definite matrix B, that is $B > T$ or $B - T > O$, then (5.1.4) is Riemann–Liouville fractional integral of the second kind of order α, otherwise, if there is no upper bound specified, then (5.1.4) is the corresponding Weyl fractional integral. Since we are confining ourselves to real positive definite matrices, the lower bound possible for T in (5.1.3) is the null matrix O. When $A = O$ then the Riemann–Liouville fractional integral of the first kind with the lower bound O and the corresponding Weyl integrals agree here. Later the theory of fractional calculus was extended to complex domain (see [4–6]). A summary of the development in the real case will be given here, the details may be seen from the papers mentioned above. This author has also established a connection between fractional calculus and statistical distribution theory, in the scalar variable case and in real and complex matrix-variate cases. It is actually easier to interpret fractional integrals in terms of statistical densities. This aspect will be discussed in the following section.

5.2. Matrix-Variate Statistical Distributions

Let $X_1 > O, X_2 > O$ be $p \times p$ real positive definite matrix random variables and let $X_1^{\frac{1}{2}}$ and $X_2^{\frac{1}{2}}$ denote the unique positive definite square roots of X_1 and X_2, respectively. All matrices appearing in this chapter are $p \times p$ real positive definite unless stated otherwise. Let

$$U_2 = X_2^{\frac{1}{2}} X_1 X_2^{\frac{1}{2}} \quad \text{and} \quad U_1 = X_2^{\frac{1}{2}} X_1^{-1} X_2^{\frac{1}{2}}, \tag{5.2.1}$$

or $X_1 = V^{-\frac{1}{2}} U_2 V^{-\frac{1}{2}}$ and $X_1 = V^{\frac{1}{2}} U_1^{-1} V^{\frac{1}{2}}$, with $X_2 = V$ be the symmetric product and symmetric ratio of these matrices, respectively. Then from Result 2.2.5 and Problem 2.3.7 in Chapter 2 we have the Jacobians in the following format:

$$dX_1 \wedge dX_2 = |V|^{-\frac{p+1}{2}} dU_2 \wedge dV, dX_1 \wedge dX_2$$

$$= |U_1|^{-(p+1)} |V|^{\frac{p+1}{2}} dU_1 \wedge dV. \tag{5.2.2}$$

Let X_1 and X_2 be statistically independently distributed with densities $f_1(X_1)$ and $f_2(X_2)$, respectively. $[f(Y)$, where Y is an $m \times n$ matrix, is called a density when $f(Y)$ is a real-valued scalar function of Y such that $f(Y) \geq 0$ for all Y and $\int_Y f(Y) dY = 1$ where dY is the wedge product of all differentials dy_{ij}'s, where $Y = (y_{ij})$ is a matrix in the real or complex domain. Here we consider only real matrices.] Let the marginal densities of U_2 and U_1 be denoted by $g_2(U_2)$ and $g_1(U_1)$, respectively. Due to statistical independence, the joint density of X_1 and X_2 is $f_1(X_1) f_2(X_2)$. If the joint density of U_2 and V is denoted by $g(U_2, V)$ then from the Jacobian in (5.2.2)

$$g(U_2, V) = |V|^{-\frac{p+1}{2}} f_1(V^{-\frac{1}{2}} U_2 V^{-\frac{1}{2}}) f_2(V).$$

Then the marginal density of U_2, denoted by $g_2(U_2)$, is available by integrating out V from $g(U_2, V)$. That is,

$$g_2(U_2) = \int_{V > O} |V|^{-\frac{p+1}{2}} f_1(V^{-\frac{1}{2}} U_2 V^{-\frac{1}{2}}) f_2(V) dV, \tag{5.2.3}$$

and similarly the marginal density of U_1 is available by integrating out V from the joint density of U_1 and V. Then

$$g_1(U_1) = \int_{V>O} |V|^{\frac{p+1}{2}} |U_1|^{-(p+1)} f_1(V^{\frac{1}{2}} U_1^{-1} V^{\frac{1}{2}}) f_2(V) dV. \quad (5.2.4)$$

If f_1 and f_2 are real matrix-variate gamma densities then (5.2.3) gives many interesting results, connected with matrix-variate version of Krätzel integral in applied analysis, inverse Gaussian density in stochastic processes and reaction-rate probability integrals in nuclear reaction-rate theory. When f_1 is a type-1 beta density then we have connection to fractional integrals. This connection is the one that we will explore next.

5.3. Right-Sided or Second Kind Fractional Integrals

Let $f_1(X_1)$ be a real matrix-variate type-1 beta density (see Chapter 2, Eq. (2.4.10)) with parameters $\gamma + \frac{p+1}{2}$ and α or with the density

$$f_1(X_1) = \frac{\Gamma_p(\alpha + \gamma + \frac{p+1}{2})}{\Gamma_p(\alpha)\Gamma_p(\gamma + \frac{p+1}{2})} |X_1|^{\gamma} |I - X_1|^{\alpha - \frac{p+1}{2}}, \quad O < X_1 < I,$$

for $\Re(\alpha) > \frac{p-1}{2}, \Re(\gamma) > -1$, and zero elsewhere, where $\Re(\cdot)$ denotes the real part of $(\cdot)$. In statistical problems the parameters are real but our discussion of fractional integrals will involve real and complex parameters and hence we will give the conditions for complex parameters. Then (5.2.3) simplifies to the following:

$$g_2(U_2) = \frac{\Gamma_p(\alpha + \gamma + \frac{p+1}{2})}{\Gamma_p(\gamma + \frac{p+1}{2})} \frac{|U_2|^{\gamma}}{\Gamma_p(\alpha)} \int_{V>U_2}$$

$$\times |V|^{-\gamma - \alpha} |V - U_2|^{\alpha - \frac{p+1}{2}} f(V) dV \quad (5.3.1)$$

for $\Re(\alpha) > \frac{p-1}{2}, \Re(\gamma) > -1$. For $p = 1$, Kober fractional integral of order α of the second kind and parameter γ is given by the following:

$$K_{2,u,\gamma}^{-\alpha} f = \frac{u^{\gamma}}{\Gamma(\alpha)} \int_{v>u} v^{-\gamma - \alpha} (v - u)^{\alpha - 1} f(v) dv,$$

and (5.3.1) corresponds to a constant multiple of Kober fractional integral of order α of the second kind in the real scalar variable case.

Hence this author has defined Kober fractional integral of order α and of the second kind with parameter γ in the real matrix-variate case and denoted it as follows:

$$K_{2,U_2,\gamma}^{-\alpha} f = \frac{|U_2|^\gamma}{\Gamma_p(\alpha)} \int_{V>U_2} |V|^{-\gamma-\alpha} |V - U_2|^{\alpha - \frac{p+1}{2}} f(V) dV, \quad (5.3.2)$$

so that the density of the product of two real positive definite matrix-variate random variables, independently distributed, where one has a type-1 beta density with parameters $(\gamma + \frac{p+1}{2}, \alpha)$ and the other has an arbitrary density, is given by

$$g_2(U_2) = \frac{\Gamma_p(\alpha + \gamma + \frac{p+1}{2})}{\Gamma_p(\gamma + \frac{p+1}{2})} K_{2,U_2,\gamma}^{-\alpha} f. \quad (5.3.3)$$

Motivated by the observation in (5.3.3), this author has given a general definition for fractional integrals of the second kind in the real matrix-variate case as follows. Let

$$f_1(X_1) = \phi_1(X_1) \frac{|I - X_1|^{\alpha - \frac{p+1}{2}}}{\Gamma_p(\alpha)} \quad \text{and} \quad f_2(X_2) = \phi_2(X_2) f(X_2),$$

$$(5.3.4)$$

where f is an arbitrary function and ϕ_1 and ϕ_2 are specified or pre-assigned functions, and $\Re(\alpha) > \frac{p-1}{2}$.

5.3.1. Second kind fractional integrals in the matrix-variate case, general definition

Note that (5.2.3) for $p = 1$ corresponds to Mellin convolution of a product of two functions. This author called (5.2.3) as M-convolution of a product, see [1]. When this author defined M-convolutions, it was only a quantity analogous to Mellin convolution of a product in the scalar case. Now, M-convolution can be given a meaningful interpretation in terms of the density of a product of two statistically independently distributed $p \times p$ positive definite matrix random variables X_1 and X_2 when $f_1(X_1)$ and $f_2(X_2)$ are densities. Note that $g_2(U_2)$ in (5.3.1) is the density of a symmetric product of the type $U_2 = X_2^{\frac{1}{2}} X_1 X_2^{\frac{1}{2}}$. Now, the M-convolution of a product, taking f_1 and f_2 in the form (5.3.4), gives the following integral, again denoted by $g_2(U_2)$ even though $g_2(U_2)$ here need not be a statistical density. Then

$$g_2(U_2) = \int_V |V|^{-\frac{p+1}{2}} \phi_1(V^{-\frac{1}{2}} U_2 V^{-\frac{1}{2}}) \frac{1}{\Gamma_p(\alpha)} |I - V^{-\frac{1}{2}} U_2 V^{-\frac{1}{2}}|^{\alpha - \frac{p+1}{2}}$$

$$\times \phi_2(V) f(V) dV. \tag{5.3.5}$$

5.3.2. Special cases

(1): Let $\phi_1(X_1) = |X_1|^\gamma, \phi_2(X_2) = 1$. Then (5.3.5) reduces to the form

$$g_2^{(1)}(U_2) = \frac{1}{\Gamma_p(\alpha)} \int_V |V|^{-\frac{p+1}{2}} |V^{-\frac{1}{2}} U_2 V^{-\frac{1}{2}}|^\gamma$$

$$\times |I - V^{-\frac{1}{2}} U_2 V^{-\frac{1}{2}}|^{\alpha - \frac{p+1}{2}} f(V) dV$$

$$= \frac{|U_2|^\gamma}{\Gamma_p(\alpha)} \int_{V > U_2} |V|^{-\gamma - \alpha} |V - U_2|^{\alpha - \frac{p+1}{2}} f(V) dV$$

$$= K_{2,U_2,\gamma}^{-\alpha} f, \quad \Re(\alpha) > \frac{p-1}{2}$$

is the Kober fractional integral of order α of the second kind with parameter γ in the real matrix-variate case.

(2): Let $\phi_1(X_1) = 1, \phi_2(X_2) = |X_2|^\alpha$. Then (5.3.5) reduces to the form

$$g_2^{(2)}(U_2) = \frac{1}{\Gamma_p(\alpha)} \int_{V > U_2} |V - U_2|^{\alpha - \frac{p+1}{2}} f(V) dV$$

which for the case $p = 1$ is Weyl fractional integral of order α of the second kind. Hence this author has called $g_2^{(2)}(U_2)$ as Weyl fractional integral of order α of the second kind in the real matrix-variate case. If V is bounded above by a constant positive definite matrix $B > O$ then $g_2^{(2)}(U_2)$ for $p = 1$ is Riemann–Liouville fractional integral of order α of the second kind. Hence this author has called $g_2^{(2)}(U_2)$ for V bounded by $B > O$ as the Riemann-fractional integral of order α of the second kind in the real matrix-variate case. Note that U_2 is positive definite and hence $U_2 > O$.

By specializing ϕ_1 and ϕ_2 one can get various fractional integrals of the second kind introduced by various authors from time to time for $p = 1$ and hence (5.3.5) for specified functions ϕ_1 and ϕ_2 will give the corresponding real matrix-variate versions of these fractional integrals of the second kind.

5.4. Left-Sided or First Kind Fractional Integrals in the Real Matrix-Variate Case

Consider the structure in (5.2.4). This for $p = 1$ is the Mellin convolution of a ratio and it is also the density of a ratio of the form $u_1 = \frac{x_2}{x_1}$ where $x_1 > 0$ and $x_2 > 0$ are real scalar statistically independently distributed random variables. This author had called the structure in (5.2.4) as M-convolution of a ratio, analogous to Mellin convolution of a ratio in the real scalar variable case [1]. Now, this M-convolution of a ratio can be given a meaningful interpretation in terms of the density of a ratio of the form $U_1 = X_2^{\frac{1}{2}} X_1^{-1} X_2^{\frac{1}{2}}$. Let $f_1(X_1)$ in (5.2.4) be a real matrix-variate type-1 beta density with parameters (γ, α) or with the density

$$f_1(X_1) = \frac{\Gamma_p(\gamma + \alpha)}{\Gamma_p(\gamma)\Gamma_p(\alpha)} |X_1|^{\gamma - \frac{p+1}{2}} |I - X_1|^{\alpha - \frac{p+1}{2}}$$

for $O < X_1 < I, \Re(\alpha) > \frac{p-1}{2}, \Re(\gamma) > \frac{p-1}{2}$ and zero elsewhere. Then $g_1(U_1)$ of (5.2.4) has the following form:

$$g_1(U_1) = \frac{\Gamma(\gamma + \alpha)}{\Gamma_p(\gamma)} \frac{1}{\Gamma_p(\alpha)} \int_V |V|^{\frac{p+1}{2}} |V^{\frac{1}{2}} U_1^{-1} V^{\frac{1}{2}}|^{\gamma - \frac{p+1}{2}}$$

$$\times |I - V^{\frac{1}{2}} U_1^{-1} V^{\frac{1}{2}}|^{\alpha - \frac{p+1}{2}} f(V) dV$$

$$= \frac{\Gamma_p(\gamma + \alpha)}{\Gamma_p(\gamma)} \frac{|U_1|^{-\gamma - \alpha}}{\Gamma_p(\alpha)} \int_{V < U_1} |V|^{\gamma} |U_1 - V|^{\alpha - \frac{p+1}{2}} f(V) dV$$

$$= \frac{\Gamma_p(\gamma + \alpha)}{\Gamma_p(\gamma)} K_{1,U_1,\gamma}^{-\alpha} f, \Re(\alpha) > \frac{p-1}{2}, \Re(\gamma) > \frac{p-1}{2}, \quad (5.4.1)$$

where $K_{1,U_1,\gamma}^{-\alpha} f$ for $p = 1$ is Kober fractional integral of order α of the first kind with parameter γ and hence this author has called (5.4.1) as Kober fractional integral of the first kind of order α and parameter γ in the real matrix-variate case. Thus, both right-sided and left-sided or second kind and first kind fractional integrals in the scalar as well as matrix-variate cases can be given interpretations as statistical densities of product and ratio, respectively.

Again, this author has given a general definition for fractional integrals of the first kind of order α in the real matrix-variate case by taking f_1 and f_2 as in (5.3.4). Then the fractional integral of the first kind of order α in the real matrix-variate case is the following, again denoted by $g_1(U_1)$ even though $g_1(U_1)$ need not be a statistical density here:

$$
g_1(U_1)
$$

$$
= \int_V |V|^{\frac{p+1}{2}} |U_1|^{-(p+1)} \phi_1(V^{\frac{1}{2}} U_1^{-1} V^{\frac{1}{2}})
$$

$$
\times \frac{1}{\Gamma_p(\alpha)} |I - V^{\frac{1}{2}} U_1^{-1} V^{\frac{1}{2}}|^{\alpha - \frac{p+1}{2}} \phi_2(V) f(V) dV, \quad \Re(\alpha) > \frac{p-1}{2}.
$$

$$
(5.4.2)
$$

Special cases

(1): Let $\phi_1(X_1) = |X_1|^{\gamma - \frac{p+1}{2}}, \phi_2(X_2) = 1$. Then (5.4.2) reduces to the following:

$$
g_1^{(1)}(U_1) = \frac{|U_1|^{-\gamma - \alpha}}{\Gamma_p(\alpha)} \int_{V < U_1} |V|^{\gamma} |U_1 - V|^{\alpha - \frac{p+1}{2}} f(V) dV
$$

$$
= K_{1,U_1,\gamma}^{-\alpha} f, \quad \Re(\alpha) > \frac{p-1}{2}, \tag{5.4.3}
$$

which is the Kober fractional integral of the first kind of order α in the real matrix-variate case.

(2): Let $\phi_1(X_1) = |X_1|^{-\alpha - \frac{p+1}{2}}, \phi_2(X_2) = |X_2|^{\alpha}$. Then (5.4.2) becomes

$$
g_1^{(2)}(U_1) = \frac{1}{\Gamma_p(\alpha)} \int_{V < U_1} |U_1 - V|^{\alpha - \frac{p+1}{2}} f(V) dV. \tag{5.4.4}
$$

This is Weyl fractional integral of the first kind of order α for $p = 1$. Hence this author has called $g_1^{(2)}(U_1)$ as Weyl fractional integral of order α of the first kind in the real matrix-variate case. If V is bounded below by a positive definite constant matrix $A > O, A < V$ then $g_1^{(2)}(U_1)$ is Riemann–Liouville fractional integral of the first kind of order α in the real matrix-variate case. Note that, in general, $V > O$.

The various fractional integrals of the first kind, introduced by various authors from time to time for $p = 1$ or in the real scalar variable case can be extended to the corresponding real matrix-variate case by specifying ϕ_1 and ϕ_2 in (5.4.2).

By taking the series form of the hypergeometric function of matrix argument, see [1] for details, one can extend Saigo fractional integrals to the corresponding matrix-variate cases. Saigo fractional integral operators are based on Gauss' hypergeometric series $_2F_1$ for the real scalar variable case. Here there will be problems with convergence of the series when it comes to Mellin transform or Laplace transform. Hence we may take a general $_rF_s$ with $s \geq r$ to avoid such complications. In this case the argument matrix can be X_1 or $I - X_1$. That is, for $s \geq r$,

$$\phi_1(X_1) = \sum_{k=0}^{\infty} \sum_{K} \frac{(a_1)_K \dots (a_r)_K}{(b_1)_K \dots (b_s)_K} \frac{C_K(X_1)}{k!} \tag{5.4.5}$$

for $O < X_1 < I$ and zero elsewhere, or

$$\phi_1(X_1) = \sum_{k=0}^{\infty} \sum_{K} \frac{(a_1)_K \dots (a_r)_K}{(b_1)_K \dots (b_s)_K} \frac{C_K(I - X_1)}{k!} \tag{5.4.6}$$

for $O < X_1 < I$ and zero elsewhere, where $K = (k_1, \dots, k_p), k_1 + \dots + k_p = k$ and $C_K(A)$ is the zonal polynomial of order k for the matrix A. The notation $(a)_K$ stands for the following generalized Pochhammer symbol:

$$(a)_K = \prod_{j=1}^{p} \left(a - \frac{j-1}{2} \right)_{k_j}, \quad (b)_{k_j} = b(b+1) \dots$$

$$(b + k_j - 1), \quad (b)_0 = 1, b \neq 0. \tag{5.4.7}$$

Then following through the derivations of the results on Saigo operators in the real scalar case, $(p = 1)$, one can extend all the results on Saigo operators to the real matrix-variate cases. In the complex domain, the expressions for the general hypergeometric function of matrix argument, corresponding to (5.4.5) and (5.4.6), will be different, the details may be seen [1].

5.5. Fractional Derivatives in the Matrix-Variate Case

In the real scalar variable case tractional derivatives are not local activities. An integer order derivative, in the real scalar variable case, is the instantaneous rate of change at a given point and this is a local activity. But fractional derivatives are defined in terms of fractional integrals or they are certain fractional integrals and thus describe global activity, not confined to a point. Observe that Riemann integrals for the real scalar variable case cover intervals or describe activities over an interval rather than activities at a given point. When a derivative is defined as a certain integral then such a derivative can describe what is happening over an interval in the real scalar case, including what is happening at every point on that interval. This aspect makes fractional derivatives very important for practical applications. A solution coming from an integer-order differential equation gives the behavior of the function at a point. But practical situations may be somewhere near the ideal point. The ideal point and its neighborhoods are captured by the solution coming from a fractional version of the same integer order differential equation. This makes fractional-order differential equations more relevant to practical problems compared to the corresponding integer order differential equations.

Let us denote $D^{-\alpha}$ the αth-order fractional integral and D^{α} the αth-order fractional derivative. Let n be a positive integer, $n = 1, 2, \ldots$ Let $\Re(n - \alpha) > 0$. The smallest such n is $m = [\Re(\alpha)] + 1$ where $[(\cdot)]$ denotes the integer part of $(\cdot)$. The $(n - \alpha)$th-order fractional integral can be denoted as $D^{-(n-\alpha)}$. Suppose that $D = \frac{d}{dx}$ be the integer-order derivative and D^{n} the nth integer-order derivative in the real scalar variable case. Then, symbolically, one can write

$$D^{\alpha} f = D^{n} D^{-(n-\alpha)} f \quad \text{or} \quad D^{\alpha} f = D^{-(n-\alpha)} D^{n} f. \qquad (5.5.1)$$

The first form is where the $(n-\alpha)$th-order fractional integral is taken first and then it is differentiated n times. But in the second case the function f is differentiated n times first and then its $(n-\alpha)$th-order fractional integral is taken. The first process $D^n D^{-(n-\alpha)} f$ is called αth-order fractional derivative in the Riemann–Liouville sense and the second process $D^{-(n-\alpha)} D^n f$, where f is differentiated n times first and then the $(n-\alpha)$th-order fractional integral is taken, is called αth-order fractional derivative in the Caputo sense.

But when it comes to the matrix-variate case one has to define a differential operator D for the matrix-variate case first. Let $X = (x_{ij}) = X'$ be $p \times p$ where x_{ij}'s are distinct real scalar variables, except for symmetry when symmetric or when $X = X'$. Let $\frac{\partial}{\partial X} = \left(\frac{\partial}{\partial x_{ij}}\right)$ and let $\tilde{D} = |\frac{\partial}{\partial X}| =$ determinant of the matrix of partial differential operators. For example, when $X = X'$ (symmetric) and 2×2 then

$$\tilde{D} = \left|\frac{\partial}{\partial X}\right| = \begin{vmatrix} \dfrac{\partial}{\partial x_{11}} & \dfrac{\partial}{\partial x_{12}} \\[2mm] \dfrac{\partial}{\partial x_{12}} & \dfrac{\partial}{\partial x_{22}} \end{vmatrix} = \frac{\partial}{\partial x_{11}}\frac{\partial}{\partial x_{22}} - \left(\frac{\partial}{\partial x_{12}}\right)^2.$$

Let $T = (t_{ij}) = T'$ be a $p \times p$ symmetric matrix of distinct t_{ij}'s, except for symmetry. Then

$$\mathrm{tr}(TX) = \mathrm{tr}(XT) = \sum_{j=1}^{p} x_{jj}t_{jj} + 2\sum_{i>j} t_{ij}x_{ij}.$$

But if we consider a $T^* = (t_{ij}^*)$ where $t_{ij}^* = t_{ij}$ when $i = j$ and $\frac{1}{2}t_{ij}$ when $i \neq j$ or the non-diagonal elements are weighted by $\frac{1}{2}$, or if either X or T has this property, then

$$\mathrm{tr}(T^* X) = \sum_{j=1}^{p} x_{jj}t_{jj} + \sum_{i>j} x_{ij}t_{ij} = \sum_{i,j=1}^{p} t_{ij}x_{ij}.$$

Let us see what happens if $\tilde{D}$ operates on $e^{\mathrm{tr}(T^* X)}$. In order to see it clearly let us evaluate it for a 2×2 case. Here

$$\mathrm{tr}(T^* X) = t_{11}x_{11} + t_{22}x_{22} + t_{12}x_{12}.$$

Then

$$\frac{\partial}{\partial x_{11}} e^{\text{tr}(T^*X)} = t_{11} e^{\text{tr}(T^*X)}, \frac{\partial}{\partial x_{22}} e^{\text{tr}(T^*X)}$$

$$= t_{22} e^{\text{tr}(T^*X)}, \frac{\partial}{\partial x_{12}} e^{\text{tr}(T^*X)} = t_{12} e^{\text{tr}(T^*X)}.$$

Therefore

$$\frac{\partial}{\partial x_{11}} \frac{\partial}{\partial x_{22}} e^{\text{tr}(T^*X)} = t_{22} \frac{\partial}{\partial x_{11}} e^{\text{tr}(T^*X)} = t_{11} t_{22} e^{\text{tr}(T^*X)}.$$

That is

$$\left[\frac{\partial}{\partial x_{11}} \frac{\partial}{\partial x_{22}} - \left(\frac{\partial}{\partial x_{12}} \right)^2 \right] e^{\text{tr}(T^*X)} = [t_{11}t_{22} - t_{12}^2] e^{\text{tr}(T^*X)} = |T| e^{\text{tr}(T^*X)}.$$

Then the operator $\tilde{D}$ operating on $e^{\text{tr}(T^*X)}$ gives

$$\tilde{D} e^{\text{tr}(T^*X)} = |T| e^{\text{tr}(T^*X)} \text{ or } \tilde{D}^n e^{\text{tr}(T^*X)} = |T|^n e^{\text{tr}(T^*X)}. \qquad (5.5.2)$$

If the function is $e^{-\text{tr}(T^*X)}$ then

$$\tilde{D}^n e^{-\text{tr}(T^*X)} = (-1)^{np} |T|^n e^{-\text{tr}(T^*X)}. \qquad (5.5.3)$$

Let us denote for convenience $\tilde{D}_1 = \tilde{D}$ and $\tilde{D}_2 = (-1)^p \tilde{D}$ respectively so that

$$\tilde{D}_1^n e^{\text{tr}(T^*X)} = |T|^n e^{\text{tr}(T^*X)}, \quad \tilde{D}_2^n e^{-\text{tr}(T^*X)} = |T|^n e^{-\text{tr}(T^*X)}. \quad (5.5.4)$$

We can now use (5.5.4) to derive some interesting results. From the matrix-variate gamma density in Chapter 2, see (2.4.4), we have the following identity:

$$|B^*|^{-\alpha} \equiv \frac{1}{\Gamma_p(\alpha)} \int_{X>O} |X|^{\alpha - \frac{p+1}{2}} e^{-\text{tr}(B^*X)} dX \qquad (5.5.5)$$

for $X > O, B^* > O, \Re(\alpha) > \frac{p-1}{2}$. Let B^* be of the type T^* in (5.5.2) then from (5.5.4) we have the following:

$$\tilde{D}_2^n |B^*|^{-\alpha} = \frac{1}{\Gamma_p(\alpha)} \int_{X>O} |X|^{\alpha-\frac{p+1}{2}} \tilde{D}_2^n e^{-\text{tr}(B^*X)} dX$$

$$= \frac{1}{\Gamma_p(\alpha)} \int_{X>O} |X|^{\alpha-\frac{p+1}{2}} |X|^n e^{-\text{tr}(B^*X)} dX$$

$$= \frac{1}{\Gamma_p(\alpha)} \int_{X>O} |X|^{\alpha+n-\frac{p+1}{2}} e^{-\text{tr}(B^*X)} dX$$

$$= \frac{1}{\Gamma_p(\alpha)} \Gamma_p(\alpha+n) |B^*|^{-(\alpha+n)}.$$

Hence we have the following result:

Result 5.5.1.

$$\tilde{D}_2^n |B^*|^{-\alpha} = \frac{\Gamma_p(\alpha+n)}{\Gamma_p(\alpha)} |B^*|^{-(\alpha+n)}, \quad \Re(\alpha) > \frac{p-1}{2}, n = 1, 2, \ldots.$$

We can also have a similar result if the exponent of $|B^*|$ is $+\alpha$ instead of $-\alpha$. Note that (5.5.5) can be looked upon as the Laplace transform of $\frac{|X|^{\alpha-\frac{p+1}{2}}}{\Gamma_p(\alpha)}$ with the Laplace parameter matrix B^* or the right side can be written as $L_f(B^*), f = \frac{|X|^{\alpha-\frac{p+1}{2}}}{\Gamma_p(\alpha)}$. Then the inverse Laplace transform is

$$\frac{|X|^{\alpha-\frac{p+1}{2}}}{\Gamma_p(\alpha)} = \int_{\Re(B^*)>B_0} |B^*|^{-\alpha} e^{\text{tr}(B^*X)} dB^* \quad \text{for some } B_0. \quad (5.5.6)$$

Now, consider the operator $\tilde{D}_1^n$ operating on X on the left side of (5.5.6). Then

$$\tilde{D}_1^n \frac{|X|^{\alpha-\frac{p+1}{2}}}{\Gamma_p(\alpha)} = \int_{\Re(B^*)>B_0} |B^*|^{-\alpha} \tilde{D}_1^n e^{\text{tr}(B^*X)} dB^*$$

$$= \int_{B^*} |B^*|^{-\alpha} |B^*|^n e^{\text{tr}(B^*X)} dB^*$$

$$= \int_{B^*} |B^*|^{-(\alpha-n)} e^{\text{tr}(B^*X)} dB^*.$$

Interpreting the right side as an inverse Laplace transform we have the right side $\frac{|X|^{\alpha-n-\frac{p+1}{2}}}{\Gamma_p(\alpha-n)}$ for $\Re(\alpha-n) > \frac{p-1}{2}$. Hence we have the following result.

Result 5.5.2.

$$\tilde{D}_1^n \frac{|X|^{\alpha-\frac{p+1}{2}}}{\Gamma_p(\alpha)} = \frac{|X|^{\alpha-n-\frac{p+1}{2}}}{\Gamma_p(\alpha-n)}, \quad \Re(\alpha-n) > \frac{p-1}{2}$$

or

$$\tilde{D}_1^n |X|^{\alpha-\frac{p+1}{2}} = \frac{\Gamma_p(\alpha)}{\Gamma_p(\alpha-n)} |X|^{\alpha-n-\frac{p+1}{2}}, \quad \Re(\alpha-n) > \frac{p-1}{2}.$$

$$(5.5.7)$$

With the help of (5.5.6) and (5.5.7) one can compute the fractional derivatives of order α for various classes of functions, some details may be seen from Mathai (2015). Here $\tilde{D}_1$ and $\tilde{D}_2$ are not universal operators in the sense operating on any scalar function of matrix argument. A universal differential operator for real-valued scalar functions of matrix argument is still an open problem.

In this chapter we considered only real matrices. But the corresponding theory and results available for matrices in the complex domain also, see [4, 6] and other papers by this author.

Exercise 5.5

5.5.1. Evaluate the Kober fractional integral operator in (5.3.2) if (1): $f(X) = |X|^{-\delta}$; (2): $f(X) = |X|^{\delta}$ and write down the conditions of convergence.

5.5.2. If the right side of $g_2^{(2)}(U_2)$ is written as $D^{-\alpha}f$, then show that $D^{-\alpha}D^{-\beta}f = D^{-(\alpha+\beta)}f = D^{-\beta}D^{-\alpha}f$.

5.5.3. If the right side of $g_1^{(2)}(U_1)$ is denoted as $D^{-\alpha}f$, then show that $D^{-\alpha}D^{-\beta}f = D^{-(\alpha+\beta)}f = D^{-\beta}D^{-\alpha}f$.

5.5.4. Denoting the density of $u_1 = \frac{x_2}{x_1}$, where $x_1 > 0$, has a scalar variable type-1 beta density with the parameters (γ, α) and $x_2 > 0$ has an arbitrary density $f(x_2)$, in the real scalar variable case where

x_1 and x_2 are independently distributed, as $g_1(u_1)$ then by taking the Mellin convolution of a ratio, or otherwise, show that

$$g_1(u_1) = \frac{\Gamma(\alpha + \gamma)}{\Gamma(\gamma)} \frac{1}{2\pi i} \int_{c-i\infty}^{c+i\infty} \frac{\Gamma(\gamma + 1 - s)}{\Gamma(\alpha + \gamma + 1 - s)} f^*(s) u_1^{-s} ds,$$

where $f^*(s)$ is the Mellin transform of $f(x_2)$ with Mellin parameter s.

5.5.5. If $X_1 > O$ and $X_2 > O$ are $p \times p$ real matrix-variate random variables, independently distributed, where X_1 has a type-1 beta density with the parameters (γ, α) and X_2 has an arbitrary density $f(X_2)$ and if $g_1(U_1)$ denotes the density of $U_1 = X_2^{\frac{1}{2}} X_1^{-1} X_2^{\frac{1}{2}}$ then show that the M-transform of $g_1(U_1)$ with parameter s is given by the following:

$$\int_{U_1 > O} |U_1|^{s - \frac{p+1}{2}} g_1(U_1) dU_1 = \frac{\Gamma_p(\gamma + \frac{p+1}{2} - s)}{\Gamma_p(\alpha + \gamma + \frac{p+1}{2} - s)} \frac{\Gamma_p(\gamma + \alpha)}{\Gamma_p(\gamma)}$$

and write down the existence conditions.

5.5.6. For the operator $\tilde{D}_2$ defined in (5.5.4) operating on X, show that

$$\tilde{D}_2 |A \pm X|^{-\gamma} = \frac{\Gamma_p(\gamma + n)}{\Gamma_p(\gamma)} |A \pm X|^{-(\gamma + n)}, \quad A \pm X > O,$$

$$\Re(\gamma) > \frac{p-1}{2}.$$

5.5.7. Evaluate the αth-order Weyl left-sided fractional derivative $\tilde{D}^\alpha f = \tilde{D}^n \tilde{D}^{-(n-\alpha)} f, \Re(n - \alpha) > \frac{p-1}{2}$ when $f(X) = e^{\text{tr}(X)}$, where, for example, $\tilde{D}^{-(n-\alpha)} f$ is the left-sided or first kind Weyl fractional integral of order $n - \alpha$.

5.5.8. In Exercise 5.5.7 show that

$$\tilde{D}^\alpha \tilde{D}^\beta f = \tilde{D}^{\alpha+\beta} f = \tilde{D}^\beta \tilde{D}^\alpha.$$

5.5.9. Write down the $(n - \alpha)$th-order right-sided Weyl fractional integral of order α. Then evaluate the αth-order fractional derivative of the second kind in the Riemann–Liouville sense for $f(V) = |I + V|^{-\gamma}, \Re(\gamma) > \frac{p-1}{2}$ where V is $p \times p$.

5.5.10. If $\tilde{D}^\alpha$ is the αth-order fractional derivative in Exercise 5.5.9 and for the same f show that

$$\tilde{D}^\alpha \tilde{D}^\beta f = \tilde{D}^{\alpha+\beta} f = \tilde{D}^\beta \tilde{D}^\alpha f.$$

Bibliography

[1] A.M. Mathai, *Jacobians of Matrix Transformations and Functions of Matrix Argument*, World Scientific Publishing, New York, 1997.

[2] A.M. Mathai, Fractional integrals in the matrix-variate case and connection to statistical distributions, *Integral Transforms Special Funct.* **20**(12) (2009), 871–882.

[3] A.M. Mathai, Some properties of Mittag-Leffler functions and matrix-variate analogues: A statistical perspective, *Fractional Calculus & Appl. Anal.* **13**(1) (2010), 113–132.

[4] A.M. Mathai, Fractional integral operators in the complex matrix-variate case, *Linéar Algebra Appl.* **439** (2013), 2901–2913.

[5] A.M. Mathai, Fractional integral operators involving many matrix variables, *Linear Algebra Appl.* **446** (2014), 196–215.

[6] A.M. Mathai, Fractional differential operators in the complex matrix-variate case, *Linear Algebra Appl.* **478** (2015), 200–217.

Chapter 6

Lie Theory and Special Functions*

6.1. Introduction

Group representation theory has played a unique role in the development of pure mathematics and theoretical physics. Just to give a few examples, Cartan [*Rend. Circ. Mat. Palermo* **53** (1929)] suggested the link between Lie groups and special functions and that this point of view, developed by Wigner and Vilenkin provided the framework for the successive speculation on the symmetries in physics.

Many of the classical differential equations are related to Lie theory. Since a Lie group G is a complicated nonlinear object and its Lie algebra G is just a vector space endowed with a bilinear non-associative product, it is usually vastly simpler to work with Lie algebra G. This is one source of the power of Lie theory. We establish some theorems concerning these results. We show how easily properties of higher transcendental functions can be derived from theorems concerning eigenvectors for the product of two operators defined on a Lie algebra of the endomorphism of vector space V.

The machinery constructed by Weisner and Miller using representation theory of Lie groups and Lie algebras will be applied to find a realization of the irreducible representation of a Lie algebra. Thus, all of the identities involving generalized transcendental functions will be given an explicit group-theoretic interpretation instead

---

*This chapter is summarized from the lectures given by Professor Dr M.A. Pathan.

of being considered merely as the result of some formal manipulation of infinite series.

This work deals with the study of properties of some generalized special functions using Lie algebraic and operational techniques. Generating relations involving many classes of special functions in several variables and some classes of well-known special polynomials and functions will be derived by using suitable Lie group theoretic method. Further, new families of special polynomials and functions, particularly useful in physics will be introduced and their properties will be derived by using operational methods and by using a combination of Lie algebraic and operational techniques. Furthermore, some Lie theoretic generating relations, representations, summation formulas and mixed generating relations for generalized special functions will be derived by using the methods of Weisner, Miller and other techniques. Many results will be derived related to the generalized special functions along with the relevant applications.

6.2. Matrix Groups

Exponential of a matrix and its properties are vital in the development of the concepts of classical groups, generators of groups theory and Lie algebra.

Definition 6.2.1 (Exponential of a matrix). If A be an $n \times n$ real matrix, then

$$\mathrm{e}^A = I + A + \frac{A^2}{2!} + \cdots = \sum_{n=0}^{\infty} \frac{A^m}{m!}$$

is an exponential matrix of A where A^2 means the matrix product $A \times A$.

Further e^A converges if each member of the sequence converges. This exponential behaves like e^x, $x \in \mathbb{R}$ and $\mathrm{e}^O = I$ where O is the null matrix. For any $n \times n$ matrix e^A is non-singular. First we consider exponential of a diagonal matrix. If the matrix A is diagonal like

$$A = \begin{pmatrix} 1 & 0 \\ 0 & -1 \end{pmatrix},$$

then its nth power is also a diagonal matrix with the diagonal elements raised to the nth power

$$A^n = \begin{pmatrix} 1 & 0 \\ 0 & (-1)^n \end{pmatrix}.$$

Then summing the exponential series, element by element, yields

$$e^A = \begin{pmatrix} \displaystyle\sum_{n=0}^{\infty} \frac{1}{n!} & 0 \\ 0 & \displaystyle\sum_{n=0}^{\infty} \frac{(-1)^n}{n!} \end{pmatrix} = \begin{pmatrix} e & 0 \\ 0 & \dfrac{1}{e} \end{pmatrix}.$$

Let $A = \begin{pmatrix} 0 & 1 \\ 0 & 0 \end{pmatrix}$ and $B = \begin{pmatrix} 0 & 0 \\ 1 & 0 \end{pmatrix}$. Then we find that $e^A = \begin{pmatrix} 1 & 1 \\ 0 & 1 \end{pmatrix}$ and $e^B = \begin{pmatrix} 1 & 0 \\ 1 & 1 \end{pmatrix}$. To evaluate e^{A+B}, one can use power series to compute

$$e^{A+B} = e^{B+A} = \sum_{n=0}^{\infty} \frac{(A+B)^n}{n!} = \sum_{n=0}^{\infty} \frac{(A+B)^{2n}}{2n!} + \sum_{n=0}^{\infty} \frac{(A+B)^{2n+1}}{(2n+1)!}$$

$$= \sum_{n=0}^{\infty} \frac{1}{2n!} I + \sum_{n=0}^{\infty} \frac{1}{(2n+1)!}(A+B)$$

$$= \cosh(1)I + \sinh(1)(A+B)$$

$$= \begin{pmatrix} \dfrac{e+e^{-1}}{2} & \dfrac{e-e^{-1}}{2} \\ \dfrac{e-e^{-1}}{2} & \dfrac{e+e^{-1}}{2} \end{pmatrix}.$$

Now compute $e^A e^B = \begin{pmatrix} 2 & 1 \\ 1 & 1 \end{pmatrix}$ and $e^B e^A = \begin{pmatrix} 1 & 1 \\ 1 & 2 \end{pmatrix}$.

These calculation show that e^{A+B}, $e^A e^B$ and $e^B e^A$ can all be different and warn us that familiar functional identities for scalar functions may not carry over to such functions of matrices.

Theorem 6.2.1. *Let $A, B \in M_n$ be given. If A and B commute, then $e^{A+B} = e^A e^B = e^B e^A$. In particular, $e^{-A} = (e^A)^{-1}$ and $e^{mA} = (e^A)^m$ for any $A \in M_n$ and any integer $m = \pm 1, \pm 2, \ldots$.*

Proof. Use the power series and binomial expansion to compute

$$e^A e^B = \sum_{j=0}^{\infty} \frac{A^j}{j!} \sum_{n=0}^{\infty} \frac{B^n}{n!} = \sum_{n=0}^{\infty} \frac{1}{n!} \sum_{j=0}^{n} \binom{n}{j} A^j B^{n-j}$$

$$= \sum_{n=0}^{\infty} \frac{(A+B)^n}{n!} = e^{A+B}.$$

For $B = -A$, we find that $e^{A-A} = e^0 = I$. For $B = A$, we have $e^{2A} = (e^A)^2$.

Example 6.2.1. If A is a real skew symmetric matrix then e^A is orthogonal. Let A' denote the transpose of A. For a skew symmetric matrix $A' = -A$.

$$I = e^O = e^{A-A} = e^{A+A'} = (e^A)(e^{A'}) = (e^A)(e^A)'$$

which implies that e^A is orthonormal.

Example 6.2.2. If $A = \begin{pmatrix} 0 & x \\ -x & 0 \end{pmatrix}$ be a real skew symmetric matrix, then e^A is special orthogonal group SO(2).

Using the definition of the exponential of a matrix, we can find that

$$e^A = \begin{pmatrix} \cos x & \sin x \\ -\sin x & \cos x \end{pmatrix} = 1, \quad x \in \mathbb{R},$$

which is a plane motion of x rotations. Since $e^A = 1$, e^A is called special orthogonal group SO(2).

Exercise 6.2

6.2.1. Let $A = \begin{pmatrix} a & c \\ 0 & b \end{pmatrix}$. Then show that $e^A = \begin{pmatrix} e^a & d \\ 0 & e^b \end{pmatrix}$ for any $a, b, c \in \mathbb{C}$, where $d = \frac{c(e^b - e^a)}{b-a}$ if $b \neq a$ and $d = c e^a$ if $b = a$.

6.2.2. Consider $A = \begin{pmatrix} 0 & 0 \\ 0 & 2 \end{pmatrix}$ and $B = \begin{pmatrix} 0 & 0 \\ 1 & 0 \end{pmatrix}$. Verify that A and B do not commute and that $e^A = e^B = e^{A+B} = I$. Prove that $e^A e^B = e^B e^A = e^{A+B}$.

6.2.3. Consider $A = \begin{pmatrix} \pi i & 0 \\ 0 & -\pi i \end{pmatrix}$ and $B = \begin{pmatrix} 0 & 1 \\ 0 & 0 \end{pmatrix}$. Verify that A and B do not commute and that $e^A = e^{A+B} = -I$. Prove that $e^A e^B \neq e^{A+B}$.

6.2.4. Prove that $e^{\text{tr}(A)} = \det e^{(A)}$

6.2.5. Let A be a matrix whose trace is zero. Then e^A has determinant 1.

6.3. Manifold

In this section we quickly review the basic concept of the theory of smooth manifolds. For us the word smooth (analytic) means infinitely often differentiable. Throughout this work we use the words smooth, differentiable, and $\mathcal{C}^\infty$ as synonymous. Let $\mathbb{R}^k$ be n-dimensional Euclidean space.

Definition 6.3.1 (Smooth maps). Let $U \subset \mathbb{R}^k$ and $V \subset \mathbb{R}^k$ be open sets. A mapping f from U to V (written as $f : U \to V$) is called smooth if all the partial derivatives $\frac{\partial^n f}{\partial x_1 \ldots \partial x_n}$ exist and are continuous.

Definition 6.3.2 (Curves in vector space V). By a curve v in V, we mean a continuous function $v : (a, b) \to V$ where (a, b) is an open interval in $\mathbb{R}$ and V is a finite dimensional vector space.

For c in (a, b), v is differential at c, if $\lim_{h \to 0} \frac{v(c+h)-v(c)}{h}$ exists. If this limit exists, it is a vector in V. We call it a *tangent vector* at c and denote by $v'(c)$

Definition 6.3.3 (Analytic curve). If a curve v in V can be differentiated as many times as we require, then v is an analytic curve.

Definition 6.3.4 (Diffeomorphism). A map $f : X \to Y$ is called a diffeomorphism if f carries X homeomorphically onto Y and f and f^{-1} are smooth.

Definition 6.3.5 (Smooth manifold). A subset $M \subset \mathbb{R}^k$ is a smooth manifold of dimension n, if each $x \in M$ has a neighborhood $W \cap M$ diffeomorphic to an open subset U of $\mathbb{R}^n$.

If $x \in M$ has a neighborhood $W \cap M$ consisting of x alone, then M is a zero-dimensional manifold. Any diffeomorphism $g : U \to W \cap M$ is called a *parameterization* of the region $W \cap M$ and the inverse diffeomorphism $W \cap M \to U$ is called a *system of coordinates*.

Definition 6.3.6. If G is the matrix group, then its dimension is the dimension of the vector space T of tangent vectors at the identity.

Example 6.3.1. The unit sphere S^2 consisting of all $(x, y, z) \in \mathbb{R}^3$ with $x^2 + y^2 + z^2 = 1$ is a smooth manifold of dimension 2. The diffeomorphism $(x, y) \to (x, y, \sqrt{1 - x^2 - y^2})$ for $x^2 + y^2 < 1$ parameterizes the region $z > 0$ of S^2.

6.4. Lie Groups

The distinguishing feature of a Lie group is that it also carries the structure of a smooth manifold, so the group elements can be continuously varied. An r-parameter Lie group is a group G which carries the structure of an r-dimensional smooth manifold in such a way that both the group operation $G \times G \to G$ and the inversion $G \to G$ are smooth maps between manifolds.

6.4.1. Local Lie groups

Let $\mathcal{C}^n$ be the space of complex n-tuples $g_n = (g_1, g_2, \dots, g_n)$ where $g_i \in \mathcal{C}$ for $i = 1, 2, \dots, n$ and define the origin $\mathbf{e}$ of $\mathcal{C}^n$ by $\mathbf{e} = (0, 0, 0)$. Suppose U is an open set in $\mathcal{C}^n$ containing $\mathbf{e}$.

Definition 6.4.1. *A complex n-dimensional local Lie group G in the neighborhood $U \subseteq \mathcal{C}^n$ is determined by a function $\Phi : \mathcal{C}^n \times \mathcal{C}^n \to \mathcal{C}^n$ such that*

(1) $\Phi(g, h) \in \mathcal{C}^n$ for all $g, h \in U$;
(2) $\Phi(g, h)$ is analytic in each of its $2n$ arguments;
(3) If $\Phi(g, h) \in U, \Phi(h, k) \in U$ then $\Phi(\Phi(g, h), k) = \Phi(g, \Phi(h, k))$;
(4) $\Phi(g, \mathbf{e}) = \Phi(\mathbf{e}, g) = \mathbf{e}$ for all $g \in \mathbf{e}$.

If we write gh for $\Phi(g, h)$ and g^{-1} for the multiplicative inverse of g, then the above axioms translate into usual group axioms, except that they are not necessarily defined everywhere. Thus gh makes sense only for g and h sufficiently near the origin $\mathbf{e}$. The associativity law (3) says that $g(hk) = (gh)k$ whenever both sides of the equation are defined. The identity element of the group is the origin $\mathbf{e}$ of $\mathcal{C}^n$. Finally, g^{-1} again is only defined for g sufficiently near $\mathbf{e}$, and $gg^{-1} = \mathbf{e} = g^{-1}g$ for such g's. Property (4) defines $\mathbf{e}$ as the group identity element. In particular, a neighborhood of the identity element of a global Lie group is the local Lie group.

Definition 6.4.2. A *global Lie group* G is a group that is also a smooth manifold such that the product map

$$\lambda : G \times G \to G,$$

$$\lambda : (g, h) \to gh, \quad g, h \in G$$

and inverse map

$$\mu : G \to G,$$

$$\mu : g \to g^{-1}, \quad g \in G$$

are smooth.

6.4.2. Examples of Lie groups

Example 6.4.1. A zero-dimensional Lie group is a discrete group. For then $\mathbf{e}$ has the neighborhood consisting of the identity origin $\mathbf{e}$ alone, which may be regarded as being homeomorphic to $\mathbb{R}^0 = \{\mathbf{e}\}$.

Example 6.4.2. We know that $G = \mathbb{R}$, the set of real numbers is also a group with addition serving as the group operation. 0 is the identity, and $-x$ the inverse of the real number x. In both of these cases the group operation is commutative. This is an Abelian group and is a Lie group since it is also a smooth manifold. The group structure of $\mathbb{R}^n$ is that of an n-dimensional vector space over the field $\mathbb{R}$.

Example 6.4.3. *General linear groups* $\mathrm{GL}(n, \mathbb{R})$. The set of all non-invertible $n \times n$ matrices with real entries is a group under multiplication, the identity element is the identity matrix I, the inverse of the matrix is the ordinary matrix inverse. This group is known as *general linear group* and is denoted by $\mathrm{GL}(n, \mathbb{R})$. For brevity, we denote $\mathrm{GL}(n, \mathbb{R})$ by $\mathrm{GL}(n)$. The distinctive feature of a Lie group is that it also carries the structure of a smooth manifold so that group elements can be continuously varied. Thus $\mathrm{GL}(n)$ is a Lie group since it is also a smooth manifold. As for $\mathrm{GL}(n)$, it can be identified with open subset

$$\mathrm{GL}(n) = \{X : \det(X) \neq 0\}$$

of the space of all $M_{n \times n}$ of all $n \times n$ matrices. But $M_{n \times n}$ is isomorphic to $\mathbb{R}^{n^2}$, the coordinate becomes the matrix entries x_{ij} of X. Thus $\mathrm{GL}(n)$ is also an n^2-dimensional manifold in which the group operation is smooth. Similarly, we can prove that a general linear group of all complex $n \times n$ matrices $\mathrm{GL}(n, \mathbb{C})$ is a Lie group.

Example 6.4.4. *Orthogonal groups* $O(n, \mathbb{R})$ are Lie groups because they depend on continuously varying parameters (the Euler angles and their generalization for higher dimensions) and they are compact because the angles vary over closed and finite intervals. Define the subset $O(n, \mathbb{R})$ of $\mathrm{GL}(n, \mathbb{R})$ as follows:

$$O(n, \mathbb{R}) = \{A \in \mathrm{GL}(n, \mathbb{R} : A({}^t A) = I)\},$$

where I is the identity matrix. Since $I({}^t I) = I$, the identity matrix $I \in O(n, \mathbb{R})$. If $A({}^t A) = I$, $B({}^t B) = I$, then $(AB)({}^t(AB)) = AB({}^t B)({}^t A) = I$ and therefore $AB \in O(n, \mathbb{R})$ which means that $O(n, \mathbb{R})$ is closed under multiplication. Also, if $A({}^t A) = I$ then $A^{-1} = {}^t A$ which means that $A^{-1}({}^t A^{-1}) = ({}^t A)A = A^{-1}A = I$ and hence $A \in O(n, \mathbb{R})$ implies $A^{-1} \in O(n, \mathbb{R})$. Hence $O(n, \mathbb{R})$ is a Lie subgroup of $\mathrm{GL}(n, \mathbb{R})$. This real orthogonal group is of dimension $\frac{n^2-n}{2}$ and is compact and closed submanifold of $\mathrm{GL}(n, \mathbb{R})$.

6.4.3. Special orthogonal groups $\mathrm{SO}(n, \mathbb{R})$

The set

$$\mathrm{SO}(n, \mathbb{R}) = \{A \in O(n, \mathbb{R}) : \det A = I\}$$

is a subgroup of a group $O(n, \mathbb{R})$ which is closed and hence compact subset of $O(n, \mathbb{R})$ and is called special orthogonal group. The condition $A({}^t A) = I$ implies that $(\det(A))^2 = 1$ and hence $\det(A) = \pm 1$ and therefore $\mathrm{SO}(n, \mathbb{R})$ is a subset of $O(n, \mathbb{R})$ satisfying $\det(A) = 1$. The dimension of $\mathrm{SO}(n, \mathbb{R})$ is $\frac{n^2-n}{2}$.

Remark 6.4.1. In general, orthogonal group $O(n, \mathbb{C})$ is the group of all complex matrices A such that $A({}^t A) = I$. As the reader can check, there are $\frac{n^2-n}{2}$ independent matrix elements in $n \times n$ skew symmetric matrix, so $O(n, \mathbb{C})$ is a complex Lie group of complex dimension $\frac{n^2-n}{2}$ [or, an $n(n-1)$-dimensional real Lie group] and is compact.

Remark 6.4.2. Let $G = SO(2)$ be the group of rotations in the plane. That is,

$$G = \begin{pmatrix} \cos\theta & -\sin\theta \\ \sin\theta & \cos\theta \end{pmatrix}, \quad 0 \le \theta < 2\pi, \tag{1}$$

where θ denotes the angle of rotation. Observe that we can identify G with the unit circle $S^1 = (\cos\theta, \sin\theta)$ such that $0 \le \theta < 2\pi$ is in $\mathbb{R}^2$, which defines the manifold structure on $SO(2)$.

Remark 6.4.3. We can check that the following expressions are subgroups of $GL(n, K)$:

(a) $SL(n, K) = \{A \in GL(n, K) : \det(A) = 1\}$,
(b) $GL_+(n, \mathbb{R}) = \{A \in GL(n, K) : \det(A) > 0\}$,
(c) $\Lambda(n, K) = \{A \in GL(n, K) : A\text{-diagonal matrix}\}$,
(d) $T(n, \mathbb{K}) = \{A \in GL(n, K) : A\text{-upper triangular matrix}\}$,

where by K we mean either the field $\mathbb{R}$ of real numbers or the field $\mathbb{C}$ of complex numbers.

Example 6.4.5. The group $T(n)$ of upper triangular matrices with all 1 on the main diagonal is an $\frac{n^2-n}{2}$-parameter Lie group. As a manifold, $T(n)$ can be identified with the Euclidean space $\mathbb{R}^{\frac{n^2-n}{2}}$ since each matrix is uniquely determined by its entries above the diagonal. For instance, in the case of $T(3)$, we identify the matrix

$$\begin{pmatrix} 1 & x & z \\ 0 & 1 & y \\ 0 & 0 & 1 \end{pmatrix} \in T(3)$$

with the vector (x, y, z) in $\mathbb{R}^3$. However, except in the special case of $T(2)$, $T(n)$ is not isomorphic to the Abelian Lie group $\mathbb{R}^{\frac{n^2-n}{2}}$. In the case of $T(3)$, the group operation is given by

$$(x, y, z)(x', y', z') = (x + x', y + y', z + z' + xy')$$

using the above identification. This is not the same as vector addition, in particular, it is not commutative. Thus a fixed manifold may be given the structure of a Lie group in more than one way.

Example 6.4.6.

$$E(2) = \left\{ \begin{pmatrix} \cos\theta & -\sin\theta & 0 \\ \sin\theta & \cos\theta & 0 \\ a & b & 1 \end{pmatrix} : a, b \in \mathbb{R}, 0 \leq \theta < 2\pi \right\}$$

is a group with respect to multiplication. Also $E(2) = \{a, b, \theta : a, b \in \mathbb{R}$ and $0 \leq \theta < 2\pi\}$. Since the parameters a, b and θ are independent, $E(2)$ is a three-dimensional manifold. $E(2)$ satisfies the associative law given in condition (3) of the definition of Lie group. Therefore $E(2)$ is a Lie group.

Example 6.4.7. The group $GL(V)$ of invertible linear transformations of an n-dimensional vector space V of the field K (by K we mean either the field $\mathbb{R}$ of real numbers or the field $\mathbb{C}$ of complex numbers) can be regarded as a Lie group in view of the isomorphisms $GL(V) \cong GK(n, K)$, which assigns to each linear transformation its matrix with respect to some fixed basis.

Example 6.4.8. The circle $\mathbb{T} = \{z \in \mathbb{C} : |z| = 1\}$ is a real one-dimensional Lie group with respect to multiplication of complex numbers as the group operation and the usual differentiable structure as a manifold.

Example 6.4.9. *Example of a local but not global Lie group.* A non-trivial example of a local but not global one-parameter Lie group is $V = \{x : |x| < 1\} \subset \mathbb{R}$ with group multiplication

$$m(x, y) = \frac{2xy - x - y}{xy - 1}; \quad x, y \in V.$$

The associativity and identity laws for $m(x, y)$ can be verified. The inverse map is $i(x) = \frac{x}{2x-1}$ defined for

$$x \in V_0 = \left\{ x : |x| < \frac{1}{2} \right\}.$$

Therefore m is a local one-parameter Lie group.

Remark 6.4.4. The composite function $\Phi(x, y) = (x^3 + y^3)^{\frac{1}{3}}$ defines the set $\mathbb{R}$ of real numbers (with the usual topology) as a topological group, but not as a Lie group, because Φ is not analytic.

6.5. Lie Bracket and Lie Algebra

Let G be a local Lie group in the neighborhood $V \subset C^n$ and let

$$t \longrightarrow g(t) = (g_1(t), \ldots, g_n(t)), \quad t \in C,$$

be an analytic mapping of a neighborhood of $o \in C$ into V such that $g(0) = e$. Such a mapping may be taken as an analytic curve in G passing through the identity e. An *infinitesimal vector* α of an analytic curve $g(t)$ in V is the *tangent vector at e*, that is

$$\alpha = \frac{\mathrm{d}}{\mathrm{d}t} g(t)|_{t=0} = \left(\frac{\mathrm{d}}{\mathrm{d}t} g_1(t), \ldots, \frac{\mathrm{d}}{\mathrm{d}t} g_n(t) \right)\Big|_{t=0} \in C^n.$$

The analytic curve

$$g(t) = (\alpha_1 t, \ldots, \alpha_n t)$$

has the infinitesimal vector $\alpha = (\alpha_1, \ldots, \alpha_n)$. If $g(t)$, $h(t)$ are analytical curves in G such that $g(0) = h(0) = e$ and let α, β be tangent vectors of $g(t)$, $h(t)$ at e, respectively, then the analytical curve $g(t)h(t)$ has tangent vector $\alpha + \beta$ at e where plus sign refers to the usual vector addition. In this way, we connect addition of infinitesimal vectors with the group multiplication in V.

Definition 6.5.1. The commutator $[\alpha, \beta]$ of α and β is the infinitesimal vector of the analytic curve

$$k(t) = g(\tau)h(\tau)g^{-1}(\tau)h^{-1}(\tau), \quad t = \tau^2,$$

that is,

$$[\alpha, \beta] = \frac{\mathrm{d}}{\mathrm{d}t} k(t)|_{\tau=0} = \alpha\beta - \beta\alpha, \quad t = \tau^2.$$

More precisely, $[\alpha, \beta]$ is the coefficient of τ^2 in the Taylors series.

The commutator has the following properties:

$$[\alpha, \beta] = -[\beta, \alpha],$$

$$[a\alpha + b\beta, \gamma] = a[\alpha, \gamma] + b[\beta, \gamma],$$

$$[[\alpha, \beta], \gamma] + [[\beta, \gamma], \alpha] + [[\gamma, \alpha], \beta] = 0,$$

where $a, b \in C$ and $\alpha, \beta, \gamma \in C^n$.

Remark 6.5.1. Note that if h commutes with g then $k = e$. As an example, if we take $g = e^{\alpha\tau}, h = e^{\beta\tau}$ then, ignoring second-order quantities, we have $k = e^{\alpha\tau}e^{\beta\tau}e^{-\alpha\tau}e^{-\beta\tau} = (e + \alpha\tau + \cdots)(e + \beta\tau + \cdots)(e - \alpha\tau + \cdots)(e - \alpha\tau + \cdots) = e + \tau^2(\alpha\beta - \beta\alpha) + \cdots$ since $[\alpha, \alpha] = 0$. Therefore

$$[\alpha, \beta] = \frac{d}{dt}k(t)|_{\tau=0} = \alpha\beta - \beta\alpha, \quad t = \tau^2.$$

Definition 6.5.2. *A complex Lie algebra.* $\mathcal{G}$ is a complex vector space together with a multiplication $[\alpha, \beta] \in \mathcal{G}$ defined for all $\alpha, \beta \in \mathcal{G}$ such that

  (i) $[\alpha, \beta] = -[\beta, \alpha]$ (skew-symmetry);
 (ii) $[a_1\alpha_1 + a_2\alpha_2, \beta] = [a_1\alpha_1, \beta] + [a_2\alpha_2, \beta], a_1, a_2 \in \mathbb{C}, \alpha_1, \alpha_2, \beta \in \mathcal{G}$;
(iii) $[[\alpha, \beta], \gamma] + [[\beta, \gamma], \alpha] + [[\gamma, \alpha], \beta] = \Theta$ (Jacobi identity).

Properties (i) and (ii) imply that $[\alpha, \beta]$ is bilinear. The Jacobi identity may be proved by expanding each of the foregoing commutators. Note that the identity may also be written in the form

$$[[\alpha, \beta], \gamma] = [\alpha, [\beta, \gamma]] + [[\alpha, \gamma], \beta].$$

In this form it has a strong resemblance to the equation

$$\frac{d}{dx}(f(x)g(x)) = \frac{d}{dx}(f(x))g(x) + \frac{d}{dx}(g(x))f(x).$$

(Check that $\alpha \to \frac{d}{dx}, \beta \to f(x), \gamma \to g(x)$.) It is for this reason that the Lie bracket $[,]$ is sometimes called a derivative. Lie algebra of a Lie group G is traditionally denoted by the corresponding lower case German letter $\mathcal{G}$.

Definition 6.5.3. The Lie algebra $L(G)$ of the local Lie group G is the space of all infinitesimal vectors equipped with the operations of vector addition and Lie product. Clearly $L(G)$ is a complex abstract Lie algebra.

Definition 6.5.4. A *one-parameter subgroup* of a local Lie group G is an analytic curve $g(t)$ in G defined in some neighborhood U of $0 \in C$ such that

$$g(t_1)g(t_2) = g(t_1 + t_2), \quad t_1, t_2, t_1 + t_2 \in U. \tag{6.5.1}$$

Theorem 6.5.1. *For each* $\alpha \in L(G)$ *there is unique one-parameter subgroup*

$$g(t) = \exp(\alpha t) \tag{6.5.2}$$

with infinitesimal vector α. *Furthermore, every one-parameter subgroup* (6.5.2) *is of the form* (6.5.1). *In particular if* $L(G)$ *is a Lie algebra of matrices then*

$$\exp(\alpha t) = \sum_{j=0}^{\infty} \frac{t^j}{j!} \alpha^j.$$

In the exponential mapping $\alpha \longrightarrow \exp(\alpha t)$, *which is implicit in Theorem 6.5.1, we set* $t = 1$.

Theorem 6.5.2. *The exponential mapping* $\alpha \longrightarrow exp(\alpha)$ *is a one-to-one transformation of a neighborhood of* $\theta \in L(G)$ *onto a neighborhood of* $e \in G, \theta$ *being the additive identity of* $L(G)$. *Furthermore, if* $g_1, \ldots, g_n$ *form a basis for* $L(G)$ *then every element* g *in small neighborhood of* $e \in G$ *can be uniquely expressed in the form*

$$g = \exp(a_1 g_1) \exp(a_2 g_2) \ldots \exp(a_n g_n),$$

where $a_1, \ldots, a_n$ *lie in a small neighborhood of* U *of* $0 \in C$.

6.5.1. Examples of Lie algebra

Example 6.5.1. The vectors in three-dimensional vector space $\mathbb{R}^3$ becomes a Lie algebra if the bracket is defined to be the cross product

$[\bar{x}, \bar{y}] = \bar{x} \times \bar{y}$. Let $\bar{x} = (x_1, x_2, x_)$, $\bar{y} = (y_1, y_2, y_3)$ be any two vectors in $\mathbb{R}^3$. Then

$$[\bar{x}, \bar{y}] = \bar{x} \times \bar{y} = (x_2 y_3 - y_2 x_3, x_3 y_1 - x_1 y_3, x_1 y_2 - x_2 y_1)$$

and it is easily seen that

(i) skew symmetry $[\bar{x}, \bar{y}] = \bar{x} \times \bar{y}$ follows from the definition of the cross product of the vectors;

(ii) bilinearity $(\lambda \bar{x} + \mu \bar{y}) \times \bar{z} = \lambda \bar{x} \times \bar{z} + \mu \bar{y} \times \bar{z}, \lambda, \mu \in \mathbb{R}$, and $\bar{z} \times (\lambda \bar{x} + \mu \bar{y}) = \lambda \bar{z} \times \bar{x} + \mu \bar{z} \times \bar{y}$;

(iii) Jacobi identity $(\bar{x} \times \bar{y}) \times \bar{z} + (\bar{y} \times \bar{z}) \times \bar{x} + (\bar{z} \times \bar{x}) \times \bar{y} = 0$ follows by using the formula between cross product and dot product of vectors and taking into account the skew symmetry of the product of vectors.

Hence the vector space $\mathbb{R}^3$ with the operation of cross product of vectors becomes Lie algebra.

Example 6.5.2. Given any finite-dimensional vector space V, the space of endomorphisms $\operatorname{End} V = \{T : V \longrightarrow V : T \text{ is linear}\}$ equipped with the bracket $[T, S] = TS - ST$ for every $S, T \in \operatorname{End} V$ is a Lie algebra.

Example 6.5.3. *Lie algebra of a general linear group* $\operatorname{GL}(2, \mathcal{C})$. The Lie algebra of a general linear group $\operatorname{GL}(2, \mathcal{C}) = \operatorname{gl}(2)$ is the space of all complex $n \times n$ matrices with the Lie bracket $[,]$. Since each element of $\operatorname{GL}(2, \mathcal{C})$ is given by four parameters, this group is four dimensional. The special elements

$$j^+ = \begin{pmatrix} 0 & -1 \\ 0 & 0 \end{pmatrix}, \quad j^- = \begin{pmatrix} 0 & 0 \\ -1 & 0 \end{pmatrix},$$

$$j^3 = \begin{pmatrix} \frac{1}{2} & 0 \\ 0 & -\frac{1}{2} \end{pmatrix}, \quad a\varepsilon = \begin{pmatrix} 1 & 0 \\ 0 & 1 \end{pmatrix}$$

form a basis for $\operatorname{gl}(2)$ in the sense that every $\alpha \in \operatorname{gl}(2)$ can be written uniquely in the form

$$\alpha = a_1 j^+ + a_2 j^- + a_3 j^3 + a_4 \varepsilon, \quad a_1, a_2, a_3, a_4 \in \mathcal{C}$$

and this basis obeys the commutation relations

$$[j^3, j^+] = j^+, \quad [j^3, j^-] = -j^-, \quad [j^+, j^-] = 2j^3$$

and

$$[\varepsilon, j^+] = [\varepsilon, j^-] = [\varepsilon, j^3] = 0,$$

where 0 is a 2×2 matrix all of whose components are zero.

Example 6.5.4. *Lie algebra of a special linear group* SL$(2, \mathcal{C})$. The 2×2 complex special linear group SL$(2, \mathcal{C})$ is the abstract group of 2×2 non-vanishing matrices

$$g = \begin{pmatrix} a & b \\ c & d \end{pmatrix}, \quad a, b, c, d \in \mathcal{C}, ad - bc = 1.$$

Since each element of SL$(2, \mathcal{C})$ is given by three parameters, this group is three dimensional. Let us construct the tangent space to SL$(2, \mathcal{C})$ at the identity e. For this, we draw three curves through e and find the tangent vectors to these curves. If the curves are chosen such that tangent vectors to them are linearly independent, then these tangent vectors shall give the basis of the tangent vector space to SL$(2, \mathcal{C})$ at e. Let $g(t), g(0) = e$ be an analytic curve whose tangent vector at e is $\alpha = (\alpha_1, \alpha_2, \alpha_3)$. Introducing coordinates for a group element g in the neighborhood of e of SL$(2, \mathcal{C})$ in the form $g = (g_1, g_2, g_3) = (a, b, c)$, where $d = \frac{1+bc}{a}$, we have

$$\alpha = (\alpha_1, \alpha_2, \alpha_3) = \frac{\mathrm{d}}{\mathrm{d}t}(a(t), b(t), c(t))_{t=e}.$$

Now for all t, we have $a(t)d(t) - b(t)c(t) = 1$. We differentiate this identity with respect to t and put $t = 0$. As $g(0) = e$, so that $a(0) = d(0) = 1, b(0) = c(0) = 0$, we have $a'(0) + d'(0) = 0$. Thus tangent vector matrix

$$\alpha = \frac{\mathrm{d}}{\mathrm{d}t}\begin{pmatrix} a(t) & b(t) \\ c(t) & d(t) \end{pmatrix}\bigg\|_{t=0} = \begin{pmatrix} \alpha_1 & \alpha_2 \\ \alpha_3 & -\alpha_1 \end{pmatrix}$$

of SL(2) has zero trace. Thus $L[\mathrm{SL}(2)] = \mathrm{sl}(2)$ is the space of all 2×2 complex matrices with trace zero. In order to complete the proof of our assertion, we have to show that the dimension of sl(2) coincides with the dimension of the space of matrices with zero trace. Since the matrices with zero trace are linear combination of $\alpha_1, \alpha_2, \alpha_3$ and therefore the complex dimension of the space of such matrices is equal to three. Also $ad - bc = 1$ and thus each element of SL(2) is determined by three complex parameters (for example, for $a \neq 0$ by the parameters a, b, c) and so the complex dimension of this Lie algebra is also 3. Our assertion is thus completely proved: sl(2) *is the space of all complex matrices of order 2 with trace zero.* The special elements

$$j^+ = \begin{pmatrix} 0 & -1 \\ 0 & 0 \end{pmatrix}, \quad j^- = \begin{pmatrix} 0 & 0 \\ -1 & 0 \end{pmatrix}, \quad j^3 = \begin{pmatrix} \frac{1}{2} & 0 \\ 0 & -\frac{1}{2} \end{pmatrix}$$

form a basis for sl(2) in the sense that every $\alpha \in$ sl(2) can be written uniquely in the form

$$\alpha = a_1 j^+ + a_2 j^- + a_3 j^3, \quad a_1, a_2, a_3 \in \mathcal{C}$$

and these basis elements obey the commutation relations

$$[j^3, j^+] = j^+, [j^3, j^-] = -j^-, [j^+, j^-] = 2j^3.$$

For $g \in$ sl(2) and $\alpha \in$ sl(2) an integral curve $g(t) = \{g_1(t), g_2(t), g_3(t)\}$ of the system of differential equation in the matrix form can be written as

$$\frac{dg}{dt} = \alpha g, g(0) = e.$$

The function $g(t)$ is uniquely determined by $\alpha \in$ sl(2) and is the solution of this differential equation. Thus $g(t)$ is given by

$$g(t) = \exp(\alpha t) = \sum_{k=0}^{\infty} \frac{\alpha^k t^k}{k!},$$

which converges for all $t \in \mathcal{C}$.

Theorem 6.5.3. *If*

$$g = \begin{pmatrix} a & b \\ c & d \end{pmatrix} \in \mathrm{SL}(2)$$

is in a sufficiently small neighborhood of the identity, then g can uniquely be written in the form $g = \exp(b'j^+)\exp(c'j^-)$ $\exp(\tau'j^3)$ where $d = e^{-\frac{\tau'}{2}}, b' = -\frac{b}{d}, c' = -cd$ and j^+, j^-, j^3 are basis of $\mathrm{sl}(2)$.

Proof. Every $\alpha \in \mathrm{sl}(2)$ can uniquely be written in the form $\alpha = aj^+ + bj^- + cj^3, a, b, c \in \mathcal{C}$ and

$$j^+ = \begin{pmatrix} 0 & -1 \\ 0 & 0 \end{pmatrix}, \quad j^- = \begin{pmatrix} 0 & 0 \\ -1 & 0 \end{pmatrix}, \quad j^3 = \begin{pmatrix} \frac{1}{2} & 0 \\ 0 & -\frac{1}{2} \end{pmatrix}$$

satisfy

$$[j^3, j^+] = j^+, [j^3, j^-] = -j^-, [j^+, j^-] = 2j^3.$$

Then

$$\exp(aj^3) = \begin{pmatrix} e^{\frac{a}{2}} & 0 \\ 0 & e^{-\frac{a}{2}} \end{pmatrix}, \quad \exp(bj^+) = \begin{pmatrix} 1 & -b \\ 0 & 1 \end{pmatrix},$$

$$\exp(cj^-) = \begin{pmatrix} 1 & 0 \\ -c & 1 \end{pmatrix}$$

For every $g \in \mathrm{SL}(2)$, let $g = \exp(b'j^+)\exp(c'j^-)\exp(\tau'j^3)$

$$= \begin{pmatrix} e^{\frac{\tau'}{2}}(1+b'c') & -b'e^{-\frac{\tau'}{2}} \\ -c'e^{\frac{\tau'}{2}} & e^{-\frac{\tau'}{2}} \end{pmatrix} = \begin{pmatrix} a & b \\ c & d \end{pmatrix}, \quad ad - bc = 1.$$

Example 6.5.5. *Lie algebra of a local Lie group T_3.* Let τ_3 be the Lie algebra of a three-dimensional complex local Lie group T_3, a multiplicative matrix group with elements

$$g(b, c, \tau) = \begin{pmatrix} 1 & 0 & 0 & \tau \\ 0 & e^{-\tau} & 0 & \tau \\ 0 & 0 & e^{\tau} & b \\ 0 & 0 & 0 & 1 \end{pmatrix}, \quad b, c, \tau \in C.$$

T_3 has the topology of C^3 and is simply connected. A basis for τ_3 is provided by the matrices

$$j^+ = \begin{pmatrix} 0 & 0 & 0 & 0 \\ 0 & 0 & 0 & 0 \\ 0 & 0 & 0 & 1 \\ 0 & 0 & 0 & 0 \end{pmatrix}, \quad j^- = \begin{pmatrix} 0 & 0 & 0 & 0 \\ 0 & 0 & 0 & 1 \\ 0 & 0 & 0 & 0 \\ 0 & 0 & 0 & 0 \end{pmatrix},$$

$$j^3 = \begin{pmatrix} 0 & 0 & 0 & 1 \\ 0 & -1 & 0 & 0 \\ 0 & 0 & 1 & 0 \\ 0 & 0 & 0 & 0 \end{pmatrix}$$

with the commutation relations:

$$[j^3, j^{\pm}] = j^{\pm}, \quad [j^+, j^-] = 0.$$

Definition 6.5.5. Let $\mathcal{G}$ and $\mathcal{G}'$ be two Lie algebras over F. A *homomorphism* T of $\mathcal{G}$ onto $\mathcal{G}'$ is a mapping $T : \mathcal{G} \to \mathcal{G}'$ satisfying

(i) $T(\alpha x + \beta y) = \alpha T(x) + \beta T(y)$,
(ii) $T[x, y] = [T(x), T(y)]$, $x, y \in \mathcal{G}$ and $\alpha, \beta \in F$.

If T is one-to-one then homomorphism is called *isomorphism*.

Definition 6.5.6. Let T be a vector space isomorphism of a Lie algebra $\mathcal{G}$ onto a Lie algebra $\mathcal{G}'$ both over a field F. Let $j_1, j_2, \ldots$ be a basis of $\mathcal{G}$ and $T(j_i) = J_i$. Then $\mathcal{G}'$ as Lie algebra is an *isomorphic image* of $\mathcal{G}$ if $T[j_i, j_j] = [T(j_i), T(j_j)]$.

Definition 6.5.7. Let V be a vector space over F and $\mathcal{G}$ be a matrix Lie algebra over F. A *representation* of $\mathcal{G}$ on V is a homomorphism ρ of $\mathcal{G}$ into $\mathcal{L}(V)$, that is $\rho : \mathcal{G} \to \mathcal{L}(V)$ such that

(i) $\rho(\alpha x + \beta y) = \alpha \rho(x) + \beta \rho T(y)$,
(ii) $\rho[x, y] = [\rho(x), \rho(y)]$ where $x, y \in \mathcal{G}$ and $\alpha, \beta \in F$.

Definition 6.5.8. Let V be a vector space over F and let $J_1, J_2, \ldots,$ be operators in $\mathcal{L}(V)$ spanning a Lie algebra $\mathcal{G}'$. If $\mathcal{G}'$ is an isomorphic image of a matrix Lie algebra $\mathcal{G}$, then the isomorphism $\rho : \mathcal{G} \to \mathcal{G}'$ provides a representation of $\mathcal{G}$ on the representation space V.

Remark 6.5.2. (Commutator tables). We now display the structure of a given Lie algebra in tabular form. If $\mathcal{G}$ is an r-dimensional Lie algebra, and $v_1, \ldots, v_r$ form a basis for $\mathcal{G}$, then the commutator table for $\mathcal{G}$ will be the $r \times r$ table whose (i, j)th entry expresses the Lie bracket $[v_i, v_j]$. Since $[v_i, v_j] = -[v_j, v_i]$, the table is always anti-symmetric and in particular, the diagonal entries are all zero. Suppose we consider $\mathcal{G} = \mathrm{sl}(2)$, the Lie algebra of special linear group $\mathrm{SL}(2)$, which consists of all 2×2 matrices with trace 0. In this case basis elements are

$$A_1 = \begin{pmatrix} 0 & 1 \\ 0 & 0 \end{pmatrix}, \quad A_2 = \begin{pmatrix} \frac{1}{2} & 0 \\ 0 & -\frac{1}{2} \end{pmatrix}, \quad A_3 = \begin{pmatrix} 0 & 0 \\ 1 & 0 \end{pmatrix}.$$

Then we obtain the following commutator table:

$-$	A_1	A_2	A_3
A_1	0	A_1	$-2A_2$
A_2	$-A_1$	0	A_3
A_3	$2A_2$	$-A_3$	0

For example, from the table, $[A_1, A_3] = A_1 A_3 - A_3 A_1 = -2A_2$, and so on.

Proposition 6.5.1. *Let $B = \{J_1, J_2, \ldots\}$ be a basis of a vector space $\mathbb{G}$. Then $\mathbb{G}$ is a Lie algebra if each commutator $[J_i, J_j]$ is a linear combination of the vectors in the basis B.*

Example 6.5.6. Prove that the differential operators

$$J^0 = -u + z\frac{d}{dz}, \quad J^+ = -2uz + z^2 \frac{d}{dz} \quad \text{and} \quad J^- = -\frac{d}{dz}$$

generate a three-dimensional Lie algebra.

If J^0, J^+, J^- are linearly independent, then

$$\alpha J^0 + \beta J^+ + \gamma J^- = 0, \quad \alpha, \beta, \gamma \in \mathbb{R},$$

which means

$$\alpha \left(-u + z \frac{\mathrm{d}}{\mathrm{d}z} \right)$$

$$+ \beta \left(-2uz + z^2 \frac{\mathrm{d}}{\mathrm{d}z} \right) + \gamma \left(-\frac{\mathrm{d}}{\mathrm{d}z} \right) = 0 \Longrightarrow \alpha = \beta = \gamma = 0.$$

Thus we can say that J^0, J^+, J^- are independent. Therefore $B = \{J^0, J^+, J^-\}$ is a basis of a three-dimensional vector space spanned by B. Since

$$[J^0, J^+] f(z) = [J^0 J^+ - J^+ J^0] f(z)$$

$$= \left(-2uz + z^2 \frac{\mathrm{d}}{\mathrm{d}z} \right) f(z) = J^+ f(z),$$

therefore $[J^0, J^+] = J^+$. Similarly, we can prove that $[J^0, J^-] = -J^-$ and $[J^+, J^-] = 2J^0$. Hence $\{J^0, J^+, J^-\}$ generate a three-dimensional Lie algebra.

Example 6.5.7. Prove that

$$e(2) = \left\{ \begin{pmatrix} 0 & -x_1 & 0 \\ x_1 & 0 & 0 \\ x_2 & x_3 & 0 \end{pmatrix} : x_1, x_2, x_3 \in \mathbb{C} \right\}$$

is a three-dimensional complex Lie algebra.

Note that $e(2)$ is a vector space. Consider the elements

$$M = \begin{pmatrix} 0 & -1 & 0 \\ 1 & 0 & 0 \\ 0 & 0 & 0 \end{pmatrix}, \quad P_1 = \begin{pmatrix} 0 & 0 & 0 \\ 0 & 0 & 0 \\ 1 & 0 & 0 \end{pmatrix}, \quad P_2 = \begin{pmatrix} 0 & 0 & 0 \\ 0 & 0 & 0 \\ 0 & 1 & 0 \end{pmatrix}.$$

We can check that M, P_1, P_2 are linearly independent. Let

$$A = \begin{pmatrix} 0 & -x_1 & 0 \\ x_1 & 0 & 0 \\ x_2 & x_3 & 0 \end{pmatrix} \in e(2).$$

Clearly; $A = x_1 M + x_2 P_1 + x_3 P_2$ where $x_1, x_2, x_3 \in \mathbb{C}$. Therefore $\{M, P_1, P_2\}$ is a basis of $e(2)$ and

$$[M, P_1] = P_2, [M, P_2] = -P_1, [P_1, P_2] = 0,$$

0 is the linear combination of the basis elements. Hence each commutator is a linear combination of the vectors in the basis $\{M, P_1, P_2\}$. Hence it follows that $e(2)$ is a Lie algebra.

Example 6.5.8. Prove that the operators $\{\frac{\partial}{\partial x}, \frac{\partial}{\partial y}, y\frac{\partial}{\partial x} - x\frac{\partial}{\partial y}\}$ generate a Lie algebra.

6.6. Lie Algebra of the Endomorphism of the Vector Space V

Since a Lie group G is a complicated nonlinear object and its Lie algebra $\mathcal{G}$ is just a vector space endowed with a bilinear nonassociative product, it is usually vastly simpler to work with Lie algebra $\mathcal{G}$. In this section, we establish some theorems concerning these results. We show how easily properties of higher transcendental functions can be derived from theorems concerning eigenvectors for the product of two operators defined on a Lie algebra of the endomorphism of vector space.

Let $\operatorname{End} V$ be the Lie algebra of the endomorphism of the vector space V endowed with the Lie bracket $[A, B] = AB - BA$ for every $A, B \in \operatorname{End} V$. Let I be the identity operator of V.

Theorem 6.6.1. *Let $A, B \in \operatorname{End} V$ be such that $[A, B] = I$. Define the sequence $(y_n)_n \subset V$ such that $Ay_n = 0$ and $y_n = By_{(n-1)}$ for every $n \geq 1$. Then y_n is an eigenvector of eigenvalue n for BA, for every $n \geq 1$.*

Proof. First we prove that $Ay_n = ny_{n-1}$, $n \geq 1$. This is true for $n = 1$, that is, $Ay_1 = y_0$, since $[A, B] y_0 = I y_0$. Now induction is used, finally to get $BAy_n = nBy_{n-1} = ny_n$ which shows that y_n is an eigenvector of eigenvalue n for BA, for every $n \geq 1$.

Example 6.6.1. Let $V = C^\infty(R)$. Define operators $A, B \in \text{End } V$ by

$$A(f)x = \frac{1}{2}f'(x) \quad \text{and} \quad B(f)x = -f'(x) + 2xf(x)$$

for every $x \in R$, then $[A, B] = I$.

These operators give Hermite differential equation

$$\left(-\frac{1}{2}\frac{d^2}{dx^2} + x\frac{d}{dx} - n\right) y = 0, \quad y = H_n(x),$$

where Hermite polynomials $H_n(x)$ are defined by the generating function

$$e^{2xt-t^2} = \sum_{n=0}^{\infty} H_n(x)\frac{t^n}{n!}.$$

Example 6.6.2 (Rodriguez formula for Hermite polynomials).

$$B^n(1) = (-1)^n e^{x^2} \frac{d^n}{dx^n}\{e^{-x^2}\} = H_n(x).$$

Set $n = 1$ in $BAy_n = ny_n$ to get $BAy_1 = y_1$ so that $BAy_0 = y_1$. Now set $y_0 = 1$ so that $B(1) = y_1, B^2(1) = B[B(1)] = By_1 = y_2$, and so on. Finally, we get $B^n(1) = y_n$.

Hence $B^n(1) = H_n(x)$.

Example 6.6.3 (Generalized Rodriguez formula).

$$B^n f(x) = (-1)^n e^{x^2} \frac{d^n}{dx^n}\{f(x)e^{-x^2}\}.$$

For $n = 1$, we have

$$(Bf)x = -e^{x^2}\frac{d}{dx}\{e^{-x^2}\}$$

$$= -e^{x^2}\left\{\frac{df}{dx}e^{-x^2} - 2xe^{-x^2}f(x)\right\} = 2xf(x) - \frac{df}{dx},$$

which is the definition of B. Equality also holds for n. Inductively for $n+1$, we have

$$(B^{n+1}f)(x) = (-1)^{n+1}e^{x^2}\frac{\mathrm{d}^{n+1}}{\mathrm{d}x^{n+1}}\{f(x)e^{-x^2}\},$$

$$B(B^n f)(x) = 2x(B^n f)(x) - \frac{\mathrm{d}}{\mathrm{d}x}(B^n f)(x)$$

$$= 2x(-1)^n e^{x^2}\frac{\mathrm{d}^n}{\mathrm{d}x^n}\{f(x)e^{-x^2}\}$$

$$- \frac{\mathrm{d}}{\mathrm{d}x}\left\{(-1)^n e^{x^2}\frac{\mathrm{d}^n}{\mathrm{d}x^n}f(x)e^{-x^2}\right\}$$

$$= (-1)^{n+1}e^{x^2}\frac{\mathrm{d}^{n+1}}{\mathrm{d}x^{n+1}}\{f(x)e^{-x^2}\}.$$

Example 6.6.4. Using operators given in Example 6.6.1, prove that Hermite polynomials $y = H_n(x)$ satisfy the recurrence relations

(i) $H_n'(x) = 2nH_{n-1}(x)$,
(ii) $H_{n+1}(x) - 2xH_n(x) + 2nH_{n-1}(x) = 0$.

To prove (i), use $Ay_n = ny_{n-1}$. Now $A(f)x = \frac{1}{2}f'(x) \Rightarrow Ay_n = \frac{1}{2}y_n' = ny_{n-1}$.

To prove (ii), we note that $B(f)x = -f'(x) + 2xf(x)$ gives $By_n = -y_n' + 2xy_n$. Therefore $y_{n+1} = -y_n' + 2xy_n$.

Example 6.6.5. Let $V = C^\infty(\mathbb{R})$. Define operators $A, B \in \mathrm{End}\, V$ by $A(f)x = f'(x)$ and $B(f)x = xf(x)$ for every $x \in \mathbb{R}$, then $[A, B] = I$.

Using Theorem 6.6.1, we find that these operators give Laguerre differential equation:

$$x\frac{\mathrm{d}^2 y}{\mathrm{d}x^2} + (1 - x)\frac{\mathrm{d}y}{\mathrm{d}x} + ny = 0, y = L_n(x),$$

where Laguerre polynomial $L_n(x)$ is given by the generating function

$$(1 - t)^{-1}e^{\frac{xt}{t-1}} = \sum_{n=0}^{\infty} L_n(x)t^n.$$

Theorem 6.6.2. *Let $A, B \in \operatorname{End} V$ be such that $[A, B]y_n = (2n + 1)y_n$. Define the sequence $(y_n)_n \subset V$ such that $Ay_n = 0$ and $ny_n = By_{(n-1)}$ for every $n \geq 1$. Then y_n is an eigenvector of eigenvalue n^2 for BA, for every $n \geq 1$.*

Theorem 6.6.3. *Let $A, B \in \operatorname{End} V$ be such that $[A, B]y_n = (a(2n + 1) + b)y_n$, where the sequence $(y_n)_n \subset V$ is defined as follows: $Ay_1 = y_0$ and $By_n = (\frac{(a(n^2 + 2n) + bn + 1)}{an + bn + 1})y_{n+1}$ for every $n \geq 1$. Then $Ay_{n+1} = (an + bn + 1)y_n$ and y_n is an eigenvector of eigenvalue $(a(n^2 - 1) + b(n - 1) + 1)$ for BA, for every $n \geq 1$.*

Theorem 6.6.4. *Let $A, B \in \operatorname{End} V$ be such that $[A, B] = [C, D] = I$. Define the sequence $(y_{m,n})_{m,n} \subset V$ as follows: $Ay_{0,n} = Cy_{m,0} = 0$ and $By_{m-1,n} = Dy_{m,n-1} = y_{m,n}$ for every integers $m, n \geq 1$. Then $y_{m,n}$ is an eigenvector of eigenvalue $m + n$ for $(BA + DC)$, for every integer $m, n \geq 1$.*

6.6.1. Application of Theorem 6.6.4

Let $V = C^\infty(C \times C)$. Define operators $A, B, C, D \in \operatorname{End} V$ by the linear operators

$$Af(x,y) = a_1 \frac{\partial}{\partial x} + a_2 \frac{\partial}{\partial y} + a_3 f,$$

$$Bf(x,y) = b_1 \frac{\partial}{\partial x} + b_2 \frac{\partial}{\partial y} + b_3 f,$$

$$Cf(x,y) = c_1 \frac{\partial}{\partial x} + c_2 \frac{\partial}{\partial y} + c_3 f,$$

$$Df(x,y) = d_1 \frac{\partial}{\partial x} + d_2 \frac{\partial}{\partial y} + d_3 f.$$

Let $f(x, y) = \phi_{m,n} \in C^\infty(C \times C)$. Then for $[A, B] = [C, D] = I$, the identity $(BA + DC)\phi_{m,n} = (m + n)\phi_{m,n}$ holds. For special values of parameters in linear operators, we get the differential equation for two-dimensional Hermite polynomials $H_{m,n}(x, y)$

$$\left(-1/2 \frac{\partial^2}{\partial x^2} - 1/2 \frac{\partial^2}{\partial y^2} + x \frac{\partial}{\partial x} + y \frac{\partial}{\partial y} - (m + n)\right) H_{m,n}(x, y) = 0.$$

6.7. Monomiality Principle

Operational methods can be exploited to simplify the derivation of the properties associated with ordinary and generalized polynomials and to define new families of polynomials. The use of operational techniques, combined with the principle of monomiality is a fairly useful tool for treating various families of special polynomials as well as their new and known generalizations. The concept and the formalism associated with the monomiality treatment can be exploited in different ways. On one side, they can be used to study the properties of ordinary or generalized special polynomials by means of a formalism closer to that of natural monomials. On the other side, they can be useful to establish rules of operational nature, framing the special polynomials within the context of particular solutions of generalized forms of partial differential equations of evolution type.

6.7.1. Multiplicative and derivative operators

Monomiality treatment can be extended to more general cases and reduce all the problems concerning the special polynomials and eventually special functions to an abstract (yet powerful) calculus involving generalized forms of multiplicative and derivative operators $\widehat{M}$ and $\widehat{P}$, respectively. This calculus is what we currently recognize as the theory of monomiality. We recall that, according to the monomiality principle a polynomial set $\{p_n(x)\}_{n\in\mathbb{N}}$ is quasimonomial, provided there exist two operators $\widehat{M}$ and $\widehat{P}$ playing, respectively, the role of multiplicative and derivative operators, for the family of polynomials. These operators satisfy the following identities, for all $n \in N$:

$$\widehat{M}\{p_n(x)\} = p_{n+1}(x), \qquad (6.7.1)$$

$$\widehat{P}\{p_n(x)\} = p_{n-1}(x). \qquad (6.7.2)$$

The operators $\widehat{M}$ and $\widehat{P}$ also satisfy the commutation relation $[\widehat{M}, \widehat{P}] = \widehat{M}\widehat{P} - \widehat{P}\widehat{M} = \widehat{1}$ and thus display the Weyl group structure. If the considered polynomial set $p_n(x)$ is quasimonomial, its properties can easily be derived from those of the $\widehat{M}$ and $\widehat{P}$ operators. In fact (i) combining the recurrences (6.7.1) and (6.7.2), we have

$$\widehat{M}\widehat{P}\{p_n(x)\} = n\{p_n(x)\}$$

which can be interpreted as the differential equation satisfied by $\{p_n(x)\}$, if $\widehat{M}$ and $\widehat{P}$ have a differential realization. (ii) Assuming here and in the sequel $\{p_n(x)\} = 1$, then $\{p_n(x)\}$ can be explicitly constructed as

$$\{p_n(x)\} = \widehat{M}^n\{p_0(x)\} = \widehat{M}^n(1),$$

which yields the series definition for $\{p_n(x)\}$.

$$\exp(t\widehat{M})p_0(x) = \sum_{n=0}^{\infty} \frac{t^n}{n!} p_n(x).$$

6.7.2. Monomiality of Appell polynomials

The Appell polynomials are very often found in various applications in pure and applied mathematics. The Appell class contains important sequences such as the Bernoulli, Euler and Hermite polynomials and their generalized forms. The Appell polynomials may be defined by either of the following equivalent conditions: $\{A_n(x)\}_{n\in\mathbb{N}}$ is an Appell set A_n being of degree exactly n if either,

$$\text{(a)} \quad \frac{\mathrm{d}}{\mathrm{d}x}\{A_n(x)\} = nA_{n+1}(x), \quad n \in \mathbb{N},$$

or (b) there exists an exponential generating function of the form

$$\text{(i)} \quad A(t)\exp(xt) = \sum_{n=0}^{\infty} A_n(x)\frac{t^n}{n!}.$$

where $A(t)$ has (at least the formal) expansion. We recall the following result which can be viewed as an alternate definition of Appell sequences. The sequence $\{A_n(x)\}$ is Appell for $g(t)$ if and only if

$$\text{(ii)} \quad \frac{1}{g(t)}\exp(xt) = \sum_{n=0}^{\infty} A_n(x)\frac{t^n}{n!} \quad \text{where } g(t) = \sum_{n=0}^{\infty} g_n\frac{t^n}{n!}.$$

In view of (i) and (ii), we have

$$A(t) = \frac{1}{g(t)}.$$

We note that the Appell polynomials $A_n(x)$ are quasimonomial with respect to the following multiplicative and derivative operators:

$$\widehat{M_A} = x + \frac{A'(D_x)}{A(D_x)} \quad \text{or, equivalently } \widehat{M_A} = x - \frac{g'(D_x)}{g(D_x)},$$

where $\widehat{P} = D_x$ respectively.

Exercise 6.7

6.7.1. Show that for any real $n \times n$ matrix A, the sequence

$$I + A + \frac{A^2}{2!} + \cdots,$$

where I is the identity matrix, converges.

6.7.2. Let A be a matrix whose trace is zero. Then prove that e^A has determinant 1.

6.7.3. Prove that the general form of 2×2 unitary, unimodular matrix is

$$U = \begin{pmatrix} a & b \\ -b^* & a^* \end{pmatrix}$$

with $aa^* + bb^* = 1$ and a^*, b^* are complex conjugate of a and b, respectively.

6.7.4. Show that special orthogonal group SO(2) and special unitary group SU(2) are not closed under matrix multiplication.

6.7.5. Show that orthogonal groups and unitary groups are compact Lie groups.

6.7.6. Show that an $n \times n$ unitary matrix has $n^2 - 1$ independent parameters.

6.7.7. Prove that $\mathbb{R}^3$ forms a Lie algebra with Lie bracket determined by the vector cross product $[\mathbf{v}, \mathbf{w}] = \mathbf{v} \times \mathbf{w}, \mathbf{v}, \mathbf{w} \in \mathbb{R}^3$. Prove that this Lie algebra is isomorphic to so(3), the Lie algebra of the three dimensional rotation group.

6.7.8. Let l_1 and l_2 be two Abelian Lie algebras. Show that l_1 and l_2 are isomorphic if and only if they have the same dimension.

6.7.9. Suppose that $x, y \in L$ satisfy $[x, y] \neq 0$. Show that x and y are linearly independent over F.

6.7.10. Check that Jacobi identity holds.

6.7.11. Let P, Q and the identity operator I span a Lie algebra, with commutation relation $[P, Q] = I$, and of course $[P, I] = [Q, I] = 0$. [These are the canonical commutation relations.] Define $L = (P - I)QP$, and $A_n = (P - I)^n Q^n$ (so $L = A_1 P$). Then

(a) $[Q, (P - I)^n] = -n(P - I)^{n-1}$,
(b) $A_{n+1} = (A_1 + n)A_n$,
(c) $[L, A_1] = L - A_1$, and
(d) $L(A_1 + n) = (A_1 + n)L + (L + n) - (A_1 + n)$.

[Hint: Suppose ν_n is an eigenvector for L, with eigenvalue $-n$, so that $(L + n)\nu_n = 0$. Then from (d) above, $(A_1 + n)\nu_n$ is an eigenvector for L, with eigenvalue $-(n + 1)$. Conclude from (b) that if ν_0 is an eigenvector of L with eigenvalue 0 then $A_n \nu_0 = \nu_n$ is an eigenvector with eigenvalue $-n$.]

6.7.12. Let $V = C^\infty(\mathbb{R})$. Define operators $A, B \in \operatorname{End} V$ by $A(f)x = f'(x)$ and $B(f)x = xf(x)$ for every $x \in \mathbb{R}$, then prove that $[A, B] = I$ and

$$((A - I)^n f)x = e^x \frac{d^n}{dx^n}\{f(x)e^{-x}\}.$$

6.7.13. Let $J^0 = t\frac{\partial}{\partial t} + \frac{1}{2}(\alpha + 1)$, $J^+ = t[x\frac{\partial}{\partial x} + t\frac{\partial}{\partial t} + (\alpha + 1 - x)]$ and $J^- = t^{-1}[x\frac{\partial}{\partial x} - t\frac{\partial}{\partial t}]$. Prove that $\{J^0, J+, J^-\}$ generate a three-dimensional Lie algebra which is isomorphic image of $sl(2, \mathcal{C})$.

Chapter 7

Selected Topics in Multivariate Analysis*

7.1. Introduction

In the context of multivariate statistical analysis, more than one characteristic is observed on a given sampled unit. Observations involving several variables arise in every branch of the social, physical and biomedical sciences. Nowadays, an increasing number of statistical applications rely on multivariate methodologies. Certain multivariate problems are extensions of univariate ones, others arise only in several dimensions.

In this chapter, the following notation is adopted for mathematical and random quantities. Capital letters such as $X, Y, \ldots$ will denote variable matrices, either random variables or mathematical variables or values assumed by random variables. Capital letters $A, B, \ldots$ will denote constant matrices. Small letters $x, y, \ldots$ will denote real scalar mathematical variables as well as random variables. A prime will denote a transpose of a matrix. A vector random variable will be denoted by a boldfaced capital letter.

Certain properties of multivariate distributions are defined in Section 7.2, which also presents several key results on the multivariate normal distribution and quadratic forms in normal vectors. Additionally, the transformation of variables technique is described in that section. The maximum likelihood estimates of the mean value and

---

*This chapter summarizes some of the main results presented by Professor Serge B. Provost at the 2015 SERB School held at CMSS Peechi Campus, India.

covariance matrix of a normal distribution are specified in Section 7.3 where certain basic results on the Wishart distribution are provided as well. Section 7.4 introduces likelihood ratio statistics or functions thereof in connection with various tests of statistical hypotheses on the mean value vector and covariance matrix of a normal random vector, including the test statistic that enables one to carry out a multivariate analysis of variance. Hotelling's $\mathcal{T}^2$ distribution is also discussed in that section.

For the sake of brevity, the results contained in this chapter are given without motivation, proofs or illustrative examples. Their applicability will be discussed as appropriate and detailed derivations will be provided during the lectures; as well, examples will be considered in the course of the problem-solving sessions. It should be noted that certain portions of this syllabus may be omitted due to time limitations. It is assumed that the reader possesses some background in probability theory and statistical inference.

7.2. Multivariate Distributions

Continuous multivariate distributions are defined in Section 7.2.1, and some of their main properties are presented as well. The transformation of variables technique is described in Section 7.2.2. Basic results on multivariate normal vectors and quadratic forms in such vectors are presented in Sections 7.2.3 and 7.2.4, respectively.

7.2.1. Definitions and properties

If $F_{x_1,\ldots,x_p}(\alpha_1,\ldots,\alpha_p) \equiv F_X(\alpha)$ is a continuous function of the real-valued α_i's, $i = 1,\ldots,p$, and there exists a non-negative function $f_{x_1,\ldots,x_p}(x_1,\ldots,x_p) \equiv f_X(x)$ such that

$$F_{x_1,\ldots,x_p}(\alpha_1,\ldots,\alpha_p) = \int_{-\infty}^{\alpha_1} \cdots \int_{-\infty}^{\alpha_p} f_{x_1,\ldots,x_p}(x_1,\ldots,x_p)$$
$$\times \, dx_p \wedge \cdots \wedge dx_1$$

and

$$F_{x_1,\ldots,x_p}(\infty,\ldots,\infty) = 1,$$

then, $F_X(x)$ and $f_X(x)$ are respectively the cumulative distribution function ($\mathcal{CDF}$) and the probability density function ($\mathcal{PDF}$) of the continuous vector random variable $X = (x_1, \ldots, x_p)'$, the prime denoting the transpose.

For such a continuous vector random variable, one has

$$f_X(x_1, \ldots, x_p) = \frac{\partial^p F_X(x_1, \ldots, x_p)}{\partial x_1 \partial x_2 \cdots \partial x_p}.$$

Moreover

$$\mathcal{Pr}\left\{(x_1, \ldots, x_p) \in \Delta \subseteq \Re^p\right\} = \int \cdots \int_\Delta f_X(x_1, \ldots, x_p)$$
$$\times \, dx_p \wedge \cdots \wedge dx_1.$$

Let X be a p-dimensional continuous vector random variable. Then, its cumulative distribution function $F_{x_1, \ldots, x_p}(x_1, \ldots, x_p)$ possesses the following properties:

(i): $\quad \lim\limits_{\alpha_p \to \infty} F_{x_1, \ldots, x_p}(\alpha_1, \ldots, \alpha_p) = F_{x_1, \ldots, x_{p-1}}(\alpha_1, \ldots, \alpha_{p-1}).$

(ii): For each i, $i = 1, \ldots, p$,

$$\lim\limits_{\alpha_i \to -\infty} F_{x_1, \ldots, x_p}(\alpha_1, \ldots, \alpha_p) = 0.$$

(iii): $F_X(\alpha)$ is continuous to the right in each of its arguments, that is,

$$\lim\limits_{\alpha_i \to \beta_i^+} F_{x_1, \ldots, x_p}(\alpha_1, \ldots, \alpha_{i-1}, \alpha_i, \alpha_{i+1}, \ldots, \alpha_p)$$
$$= F_{x_1, \ldots, x_p}(\alpha_1, \ldots, \alpha_{i-1}, \beta_i, \alpha_{i+1}, \ldots, \alpha_p), \quad i = 1, \ldots, p.$$

(iv): $$F_{x_1, \ldots, x_p}(\infty, \infty, \ldots, \infty) = 1.$$

The components $x_1, x_2, \ldots, x_p$ of the p-variate vector random variable X are independently distributed if and only if

$$F_X(x) = \prod_{i=1}^p F_{x_i}(x_i) \quad \text{for every } x \in \Re^p$$

or, equivalently,

$$f_X(x) = \prod_{i=1}^{p} f_{x_i}(x_i) \quad \text{for every } x \in \Re^p.$$

Let $g(X)$ be a continuous function mapping $\Re^p$ onto $\Re$ and X be a p-variate continuous vector random variable whose probability density function is $f_X(X)$; then, the expected value or mathematical expectation of $g(X)$ is

$$\mathscr{E}(g(X)) = \int_{\Re} \cdots \int_{\Re} g(X) f_X(X) \, dx_1 \wedge \cdots \wedge dx_p.$$

Let $f_X(X)$ be the density function of the continuous p-variate vector random variable X. The marginal $\mathcal{PDF}$ of a subset of the components of X is obtained by integrating out from $f_X(X)$ the variables that do not belong to that subset. For example, letting $q < p$,

$$f_{x_1,\ldots,x_q}(x_1,\ldots,x_q) = \int_{\Re} \cdots \int_{\Re} f_X(x_1, x_2 \ldots, x_p)$$
$$\times \, dx_{q+1} \wedge dx_{q+2} \wedge \cdots \wedge dx_p$$

is the marginal $\mathcal{PDF}$ of $x_1, x_2, \ldots, x_q$. The conditional $\mathcal{PDF}$ of the components of a subset of X, given the complement of that subset, is equal to the $\mathcal{PDF}$ of X divided by that of the complementary subset, provided the latter is positive. For instance, the $\mathcal{PDF}$ of $x_1, \ldots, x_q$, conditional on $x_{q+1} = a_{q+1}, \ldots, x_p = a_p$, is given by

$$f_{x_1,\ldots,x_q \mid x_{q+1},\ldots,x_p}(x_1,\ldots,x_q \mid a_{q+1},\ldots,a_p)$$
$$= \frac{f_X(x_1,\ldots,x_q,a_{q+1},\ldots,a_p)}{f_{x_{q+1},\ldots,x_p}(a_{q+1},\ldots,a_p)}$$

whenever $f_{x_{q+1},\ldots,x_p}(a_{q+1},\ldots,a_p) \neq 0$.

Letting

$$\mathcal{X} = \begin{bmatrix} x_{11} & \cdots & x_{1q} \\ \vdots & \ddots & \vdots \\ x_{p1} & \cdots & x_{pq} \end{bmatrix}$$

be a $p \times q$ matrix whose elements are random variables, the mathematical expectation of this matrix random variable is

$$\mathscr{E}(\mathcal{X}) = \begin{bmatrix} \mathscr{E}(x_{11}) & \cdots & \mathscr{E}(x_{1q}) \\ \vdots & \ddots & \vdots \\ \mathscr{E}(x_{p1}) & \cdots & \mathscr{E}(x_{pq}) \end{bmatrix}.$$

For any matrix random variable $\mathcal{X}$ and any conformable constant matrices $\mathbb{A}, \mathbb{B}$ and $\mathbb{C}$, one has

$$\mathscr{E}(\mathbb{A}\,\mathcal{X}\,\mathbb{B} + \mathbb{C}) = \mathbb{A}[\mathscr{E}(\mathcal{X})]\mathbb{B} + \mathbb{C}.$$

For any vector random variable $\boldsymbol{X}$, $\mathscr{E}(\boldsymbol{X})$ is called the *mean value* or the *mean value vector* and

$$\mathscr{E}[(\boldsymbol{X} - \mathscr{E}(\boldsymbol{X}))(\boldsymbol{X} - \mathscr{E}(\boldsymbol{X}))']$$

$$= \mathscr{E}\left[\begin{pmatrix} x_1 - \mathscr{E}(x_1) \\ x_2 - \mathscr{E}(x_2) \\ \vdots \\ x_p - \mathscr{E}(x_p) \end{pmatrix} [x_1 - \mathscr{E}(x_1), x_2 - \mathscr{E}(x_2), \ldots, x_p - \mathscr{E}(x_p)]\right]$$

is referred to as the *covariance matrix* associated with $\boldsymbol{X}$, which is denoted as $\mathcal{C}ov(\boldsymbol{X})$. Thus, the (ij)th element of $\mathcal{C}ov(\boldsymbol{X})$ is $\mathscr{E}[(x_i - \mathscr{E}(x_i))(x_j - \mathscr{E}(x_j))]$. Moreover,

$$\mathcal{C}ov(\mathbb{A}\boldsymbol{X} + \boldsymbol{b}) = \mathbb{A}\,\mathcal{C}ov(\boldsymbol{X})\mathbb{A}'. \qquad (7.2.1)$$

For every real vector $\boldsymbol{T} = (t_1, t_2, \ldots, t_p)'$, the *characteristic function* of a vector random variable $\boldsymbol{X}$ is defined by $\mathscr{E}(e^{i\boldsymbol{T}'\boldsymbol{X}}) \equiv \Phi_{\boldsymbol{X}}(T)$ and its *moment-generating function* ($\mathcal{MGF}$) is

$$\mathcal{M}_{\boldsymbol{X}}(\boldsymbol{T}) = \mathscr{E}(e^{\boldsymbol{T}'\boldsymbol{X}}) = \mathscr{E}(e^{\sum_{i=1}^{p} t_i x_i}).$$

Let $\alpha_1, \ldots, \alpha_p$ be non-negative integers; then

$$\frac{\partial^{\alpha_1+\cdots+\alpha_p}\mathcal{M}_X(\mathbf{T})}{\partial t_1^{\alpha_1}\partial t_2^{\alpha_2}\ldots\partial t_p^{\alpha_p}}\Bigg|_{t_1=t_2=\cdots=t_p=0} = \mathscr{E}(x_1^{\alpha_1}x_2^{\alpha_2}\cdots x_p^{\alpha_p})$$

$$= \int_{\Re}\int_{\Re}\cdots\int_{\Re} x_1^{\alpha_1}x_2^{\alpha_2}\cdots x_p^{\alpha_p}\, f_X(\mathbf{X})\, \mathrm{d}\mathbf{X},$$

where $f_X(\mathbf{X})$ is the $\mathcal{PDF}$ of X and $\mathrm{d}\mathbf{X}$ denotes $\mathrm{d}x_1 \wedge \mathrm{d}x_2 \wedge \cdots \wedge \mathrm{d}x_p$. Let

$$\mathbf{Y}^{p\times1} = \begin{pmatrix} \mathbf{Y}_1^{r\times1} \\ \mathbf{Y}_2^{(p-r)\times1} \end{pmatrix}$$

be a continuous (or discrete) p-variate vector random variable; then, $\mathbf{Y}_1$ and $\mathbf{Y}_2$ are independently distributed subvectors if and only if

$$\mathcal{M}_Y(\mathbf{T})$$
$$= \underbrace{\mathcal{M}_Y(t_1, t_2, \ldots, t_r, 0, \ldots, 0)}_{\mathcal{MGF}\text{ of }\mathbf{Y}_1} \cdot \underbrace{\mathcal{M}_Y(0, \ldots, 0, t_{r+1}, t_{r+2}, \ldots, t_p)}_{\mathcal{MGF}\text{ of }\mathbf{Y}_2},$$

or, equivalently, if and only if $h_Y(\mathbf{Y}) = h_{Y_1}(\mathbf{Y}_1) \times h_{Y_2}(\mathbf{Y}_2)$ where $h_Y(\mathbf{Y})$ denotes the $\mathcal{PDF}$ of $\mathbf{Y}$ and $h_{Y_1}(\mathbf{Y}_1)$ and $h_{Y_2}(\mathbf{Y}_2)$ are the respective marginal $\mathcal{PDF}$'s of $\mathbf{Y}_1$ and $\mathbf{Y}_2$.

7.2.2. The transformation of variables technique

Result 7.2.1. *Let $x_1, \ldots, x_n$ be continuous real scalar random variables whose joint $\mathcal{PDF}$, $f_X(\mathbf{X}) \equiv f_{x_1,\ldots,x_n}(x_1,\ldots,x_n)$, is defined on a set $A \subseteq \Re^n$ and let $y_i = g_i(x_1,\ldots,x_n)$ for $i = 1,\ldots,n$. It is assumed that the transformation from $x_1,\ldots,x_n$ to $y_1,\ldots,y_n$ is one-to-one and that the functions $g_1(\cdot),\ldots,g_n(\cdot)$ have continuous partial derivatives with respect to $x_1,\ldots,x_n$. The Jacobian of the inverse transformation specified by*

$$x_i = h_i(y_1,\ldots,y_n), \quad i = 1,2,\ldots,n,$$

is

$$\mathscr{J}_{\mathbf{Y}\to\mathbf{X}} = \begin{vmatrix} \dfrac{\partial x_1}{\partial y_1} & \cdots & \dfrac{\partial x_1}{\partial y_n} \\ \vdots & \ddots & \vdots \\ \dfrac{\partial x_n}{\partial y_1} & \cdots & \dfrac{\partial x_n}{\partial y_n} \end{vmatrix}.$$

Let $B \subseteq \Re^n$ be the image of $A \subseteq \Re^n$ under the transformation $\mathbf{X} \to \mathbf{Y}$. Then, provided $\mathscr{J}_{\mathbf{Y}\to\mathbf{X}} \neq 0$, the joint $\mathcal{PDF}$ of the y_i's, that is, the $\mathcal{PDF}$ of the vector random variable $\mathbf{Y}_{n\times 1}$ is given by

$$f_{y_1,\ldots,y_n}(y_1,\ldots,y_n) = \begin{cases} f_{x_1,\ldots,x_n}(h_1(y_1,\ldots,y_n),\ldots,h_n(y_1,\ldots,y_n)) \\ \qquad |\mathscr{J}_{\mathbf{Y}\to\mathbf{X}}| \quad \text{if } (y_1,\ldots,y_n) \in B, \\ 0 \quad \text{otherwise.} \end{cases}$$

Note that the number of transformed variables (the y_i's) should match with the number of original variables (the x_i's). Moreover, one should ensure that the domain of the x_i's is properly mapped (for instance, one could map the boundary or the vertices of the set A to determine its image B). For example, when A covers the first quadrant of the real plane, that is, $x_1 > 0$ and $x_2 > 0$, its image under the transformation, $y_1 = x_1 + x_2$ and $y_2 = x_2$, comprises the area delimited by the lines $y_1 = y_2$ and $y_1 = 0$ within the first quadrant.

7.2.3. The multivariate normal distribution

Definition 7.2.1. *Non-singular multivariate normal distribution.* A p-variate vector random variable $\boldsymbol{X}$ is said to have a non-singular real multivariate normal distribution with mean value $\boldsymbol{\mu} \in \Re^p$ and positive definite covariance matrix $\mathbb{V}$ (that is, $\mathbb{V} \succ \mathbb{O}$) if its density function is given by

$$\eta_p(\boldsymbol{X}; \boldsymbol{\mu}, \mathbb{V}) = \frac{1}{(2\pi)^{\frac{p}{2}}|\mathbb{V}|^{\frac{1}{2}}} e^{-\frac{1}{2}(\boldsymbol{X}-\boldsymbol{\mu})'\mathbb{V}^{-1}(\boldsymbol{X}-\boldsymbol{\mu})} \quad \text{for } \boldsymbol{X} \in \Re^p.$$

This will be denoted

$$\boldsymbol{X} \sim \mathcal{N}_p(\boldsymbol{\mu}, \mathbb{V}),$$

where $\sim$ means *distributed as*. The equation $(\boldsymbol{X} - \boldsymbol{\mu})'\mathbb{V}^{-1}(\boldsymbol{X} - \boldsymbol{\mu}) = c > 0$ defines an ellipsoid in $\Re^p$ centered at the point $\boldsymbol{\mu}$.

Note that in the univariate case (letting $p = 1$, $\boldsymbol{\mu} = \mu$ and $\Sigma = \sigma^2$), the $\mathcal{PDF}$ of $x \sim \mathcal{N}(\mu, \sigma^2)$ is

$$\eta(x; \mu, \sigma^2) = \frac{e^{-\frac{1}{2}(x-\mu)'(\sigma^2)^{-1}(x-\mu)}}{(2\pi)^{\frac{1}{2}}(\sigma^2)^{\frac{1}{2}}}$$

$$= \frac{1}{\sqrt{2\pi}\,\sigma} e^{-(x-\mu)^2/(2\sigma^2)}.$$

Result 7.2.2. *Let* $\mathbf{X} \sim \mathcal{N}_p(\boldsymbol{\mu}, \mathbb{V})$. *If the covariance matrix* $\mathbb{V} = \mathcal{D}iag(\sigma_1^2, \ldots, \sigma_p^2)$, *so that* $\mathcal{C}ov(x_i, x_j) = 0$ *for* $i \neq j$, *then the components of* $\mathbf{X}$ *are independently distributed.*

Result 7.2.3. *The moment-generating function of* $\mathbf{X} \sim \mathcal{N}_p(\boldsymbol{\mu}, \mathbb{V})$, *which is defined as* $\mathscr{E}(e^{\mathbf{T}'\mathbf{X}})$, *is* $\mathcal{M}_{\mathbf{X}}(\mathbf{T}) = e^{\mathbf{T}'\boldsymbol{\mu} + \frac{1}{2}\mathbf{T}'\mathbb{V}\mathbf{T}}$.

Result 7.2.4. *The characteristic function of* $\mathbf{X}$, *that is,* $\mathscr{E}(e^{i\mathbf{T}'\mathbf{X}})$, *is then* $e^{i\mathbf{T}'\boldsymbol{\mu} - \frac{1}{2}\mathbf{T}'\mathbb{V}\mathbf{T}}$.

Result 7.2.5. *Let* $\mathbf{Y} \sim \mathcal{N}_p(\boldsymbol{\mu}, \mathbb{V})$, $\mathbb{V} \succ \mathbb{O}$, $\mathbb{A}_{r \times p}$ *be a constant matrix with* $r \leq p$ *and* $\mathbf{b}$ *is a constant* r-*dimensional vector. Then, letting* $\mathbf{Z} = \mathbb{A}_{r \times p}\mathbf{Y} + \mathbf{b}$, $\mathbf{Z} \sim \mathcal{N}_r(\mathbb{A}\boldsymbol{\mu} + \mathbf{b}, \mathbb{A}\mathbb{V}\mathbb{A}')$.

Result 7.2.6. *A characterization of the multivariate normal distribution. Let* $\mathbf{X}$ *be a vector random variable whose mean value and covariance matrix are respectively* $\boldsymbol{\mu}$ *and* $\mathbb{V}$. *If* $\mathbf{c}'\mathbf{X}$ *has a univariate normal distribution for every vector* $\mathbf{c} \in \Re^p$, *then* $\mathbf{X}$ *is distributed as a multivariate normal vector.*

Result 7.2.7. *Let* $\mathbf{X} = \begin{bmatrix} \mathbf{X}_1 \\ \mathbf{X}_2 \end{bmatrix}$ *where* $\mathbf{X}_1$ *has* q *components and* $\mathbf{X}_2$ *has* $p - q$ *components and* $\mathbf{X} \sim \mathcal{N}_p(\boldsymbol{\mu}, \mathbb{V})$ *where*

$$\boldsymbol{\mu} = \begin{bmatrix} \boldsymbol{\mu}_1^{q \times 1} \\ \boldsymbol{\mu}_2^{(p-q) \times 1} \end{bmatrix}$$

and

$$\mathbb{V} = \begin{bmatrix} \mathbb{V}_{11}^{q \times q} & \mathbb{V}_{12}^{q \times (p-q)} \\ \mathbb{V}_{21}^{(p-q) \times q} & \mathbb{V}_{22}^{(p-q) \times (p-q)} \end{bmatrix}.$$

If $\mathbb{V}_{12} = \mathbb{O}$ *and* $\mathbb{V}_{21} = \mathbb{O}$, *then* $\mathbf{X}_1$ *and* $\mathbf{X}_2$ *are independently and normally distributed with mean values* $\boldsymbol{\mu}_1$ *and* $\boldsymbol{\mu}_2$ *and covariance matrices* $\mathbb{V}_{11}$ *and* $\mathbb{V}_{22}$, *respectively.*

Result 7.2.8. *Let the normal vector random variable* $\mathbf{X}$ *be partitioned as in Result 7.2.7. Then, the conditional distribution of* $\mathbf{X}_2$, *given* $\mathbf{X}_1$, *is* $\mathcal{N}_{p-q}(\boldsymbol{\mu}_2 + \mathbb{V}_{21}\mathbb{V}_{11}^{-1}(\mathbf{X}_1 - \boldsymbol{\mu}_1), \mathbb{V}_{22} - \mathbb{V}_{21}\mathbb{V}_{11}^{-1}\mathbb{V}_{12})$ *and the marginal distribution of* $\mathbf{X}_1$ *is* $\mathcal{N}_q(\boldsymbol{\mu}_1, \mathbb{V}_{11})$.

7.2.4. Quadratic forms in normal vector random variables

Definition 7.2.2. Let $\mathbf{X} \sim \mathcal{N}_p(\boldsymbol{\mu}, \mathbb{V})$ and $\mathbb{A} = (a_{ij})$ be a constant symmetric matrix. The expression

$$Q(\mathbf{X}) = \mathbf{X}'\mathbb{A}\mathbf{X}$$

is said to be a *quadratic form* in the normal vector random variable $\mathbf{X}$ (or in the normal random variables $x_1, \ldots, x_p$).

Definition 7.2.3. A *quadratic expression* or *second degree polynomial* in the normal vector $\mathbf{X}$ (or in the normal variables $x_1, \ldots, x_p$) has the following representation:

$$Q^*(\mathbf{X}) = \mathbf{X}'\mathbb{A}\mathbf{X} + \mathbf{a}'\mathbf{X} + d,$$

where $\mathbb{A}$ is a constant symmetric matrix, $\mathbf{a}$ is a constant vector and d is a scalar constant.

Result 7.2.9. *Let* $\mathbf{X} \sim \mathcal{N}_p(\boldsymbol{\mu}, \mathbb{V})$ *and let* $\mathbb{A}$ *be a constant symmetric matrix. Then, the quadratic form* $\mathbf{X}'\mathbb{A}\mathbf{X}$ *can be expressed as follows in terms of independently distributed standard normal variables:*

$$\mathbf{X}'\mathbb{A}\mathbf{X} \simeq \sum_{j=1}^{p} \lambda_j (U_j + b_j)^2, \tag{7.2.2}$$

where $\lambda_1, \ldots, \lambda_p$ *are the eigenvalues of* $\mathbb{V}^{\frac{1}{2}}A\mathbb{V}^{\frac{1}{2}}$,

$$u_j \overset{\text{ind}}{\sim} \mathcal{N}(0,1), \; j = 1, \ldots, p,$$

and

$$\mathbf{b}' = (b_1, \ldots, b_p) = (\mathbb{P}'\mathbb{V}^{-\frac{1}{2}}\boldsymbol{\mu})'$$

$\mathbb{P}$ *being an orthonormal matrix* $(\mathbb{PP}' = \mathbb{I}, \mathbb{P}'\mathbb{P} = \mathbb{I})$ *such that*

$$\mathbb{P}'\mathbb{V}^{\frac{1}{2}}\mathbb{A}\mathbb{V}^{\frac{1}{2}}\mathbb{P} = \mathcal{D}\mathrm{iag}(\lambda_1, \ldots, \lambda_p).$$

Clearly, $\boldsymbol{b} = \boldsymbol{0}$ whenever $\boldsymbol{\mu} = \boldsymbol{0}$ and then

$$\mathbf{X}'\mathbb{A}\mathbf{X} \simeq \sum_{j=1}^{p} \lambda_j \, u_j^2. \tag{7.2.3}$$

Result 7.2.10. *Let* $\mathbf{X} \sim \mathcal{N}_p(\mathbf{0}, \mathbb{I})$; *then* $\mathbf{X}'\mathbb{B}\mathbf{X} \sim \chi_r^2$ *if and only if* $\mathbb{B} = \mathbb{B}^2$ ($\mathbb{B}$ *is idempotent) and* $\rho(\mathbb{B}) = r$, *that is, the rank of* $\mathbb{B}$ *is* r.

Result 7.2.11. *Craig's theorem. Let* $\mathbf{X} \sim \mathcal{N}_p(\mathbf{0}, \mathbb{I})$, $\mathbb{A} = \mathbb{A}'$ *and* $\mathbb{B} = \mathbb{B}'$; *then, the quadratic forms* $Q_1 = \mathbf{X}'\mathbb{A}\mathbf{X}$ *and* $Q_2 = \mathbf{X}'\mathbb{B}\mathbf{X}$ *are independently distributed if and only if* $\mathbb{A}\mathbb{B} = \mathbb{O}$.

Exercise 7.2

7.2.1. Establish Eq. (7.2.1).

7.2.2. Let x and y be independently distributed real scalar random variables with respective density functions $f(x) = 1$ for $0 \le x \le 1$ and $f(x) = 0$, otherwise, and $g(y) = 1$ for $0 \le y \le 1$ and $g(y) = 0$, otherwise. Consider the transformation of variables, $w = x/y$ and $z = y$. (a) Define the domain of (w, z). (b) Specify the inverse transformation and its Jacobian. (c) Obtain the joint density function of w and z. (d) Determine the density function of x/y and show that it integrates to one.

7.2.3. Show that a non-singular normal vector $\mathbf{X} \sim \mathcal{N}_p(\boldsymbol{\mu}, \mathbb{V})$ can be expressed as $\mathbb{V}^{1/2}\mathbf{Z} + \boldsymbol{\mu}$ where $\mathbf{Z} \sim \mathcal{N}_p(\mathbf{0}, \mathbb{I})$ and $\mathbb{V}^{1/2}$ denotes the positive definite square root of the real positive definite matrix $\mathbb{V}$. Then, define the standardizing transformation on $\mathbf{X}$ that results in the standardized normal vector $\mathbf{Z}$.

7.2.4. Express $\mathbf{X} \sim \mathcal{N}_p(\boldsymbol{\mu}, \mathbb{V})$ in terms of a standard normal vector $\mathbf{Z} \sim \mathcal{N}_p(\mathbf{0}, \mathbb{I})$ and apply the spectral decomposition theorem to derive Result 7.2.9.

7.2.5. If $\mathbf{W} = \alpha \mathbf{U} + \beta \mathbf{Z}$ where $\mathbf{U}$ and $\mathbf{Z}$ are multivariate normal vectors and α and β are real scalar constants, explain why $\mathbf{W}$ is also a multivariate normal vector.

7.2.6. Prove Result 7.2.7.

7.2.7. Let $\begin{bmatrix} x_1 \\ x_2 \end{bmatrix} = \mathbf{X} \sim \mathcal{N}_p(\boldsymbol{\mu}, \mathbb{V})$ where $\boldsymbol{\mu} = \begin{bmatrix} 1 \\ 4 \end{bmatrix}$ and $\mathbb{V} = \begin{bmatrix} 2 & 1 \\ 1 & 2 \end{bmatrix}$.

(a) Identify the conditional distribution of $x_1 | x_2 = 0$.
(b) Letting $y_1 = x_1 + x_2$ and $y_2 = 3 + x_2$, determine the joint density of y_1 and y_2.
(c) Obtain the expected value of x_2^2 by differentiation of the moment-generating function of $\mathbf{X}$.

7.2.8. Let x and y be independently distributed random variables with respective probability density functions $f(x) = e^{-x}$ for $x > 0$ and $f(x) = 0$, otherwise, and $f(y) = e^{-y}$ for $y > 0$ and $f(y) = 0$, otherwise. Let $w = x - y$ and $z = x$. (a) Define the domain of w and z. (b) Specify the inverse transformation and obtain its Jacobian. (c) Determine the joint density function of w and z. (d) Obtain the density function of $x - y$ and show that it integrates to one.

7.3. Maximum Likelihood Estimation

The maximum likelihood estimates of the mean value $\boldsymbol{\mu}$ and the covariance matrix $\mathbb{V}$ are obtained in Section 7.3.1 on the basis of a simple random sample of size n from a multivariate normal population. Some of the distributional properties of the corresponding estimators are enumerated in Section 7.3.2, and the Wishart distribution, which is related to the distribution of the maximum likelihood estimator of $\mathbb{V}$, is discussed in Section 7.3.3.

7.3.1. The maximum likelihood estimates of $\boldsymbol{\mu}$ and $\mathbb{V}$

Result 7.3.1. *Let* $\mathbf{X}_i = (x_{1i}, x_{2i}, \ldots, x_{pi})'$, $i = 1, 2, \ldots, n$, *be a simple random sample of size n from a multivariate normal population* *with mean value $\boldsymbol{\mu}$ and positive definite covariance matrix $\mathbb{V}$, and let*

$$\overline{\mathbf{X}} = \frac{1}{n} \sum_{i=1}^{n} \mathbf{X}_i.$$

Then, on expressing the likelihood function, that is, the product of the respective density functions of the $\mathbf{X}_i$'s, we have

$$\mathcal{L}(\boldsymbol{\mu}, \mathbb{V}) = (2\pi)^{-np/2} |\mathbb{V}|^{-n/2} \, e^{-\frac{1}{2}\text{tr}\{\mathbb{V}^{-1} \sum_{i=1}^{n}(\mathbf{X}_i - \boldsymbol{\mu})(\mathbf{X}_i - \boldsymbol{\mu})'\}}$$

$$= (2\pi)^{-np/2} |\mathbb{V}|^{-n/2} \, e^{-\frac{1}{2}\text{tr}\{\mathbb{V}^{-1}(\mathbb{S} + n(\overline{\mathbf{X}} - \boldsymbol{\mu})(\overline{\mathbf{X}} - \boldsymbol{\mu})')\}} \qquad (7.3.1)$$

where $\mathbb{S}$ is as defined in Result 7.3.3.

It can be shown that it is maximized with respect to $\boldsymbol{\mu}$ when $\boldsymbol{\mu} = \overline{\mathbf{X}}$. Thus, the maximum likelihood estimate ($\mathcal{MLE}$) of $\boldsymbol{\mu}$ is $\hat{\boldsymbol{\mu}} = \overline{\mathbf{X}}$.

The following lemma is needed to derive the maximum likelihood estimate of the covariance matrix $\mathbb{V}$.

Result 7.3.2. *Let $\mathbb{A}$ be any $p \times p$ real positive definite matrix and $f(\mathbb{A}) = c|\mathbb{A}|^{n/2} e^{-\frac{1}{2}\text{tr}(\mathbb{A})}$ where c is a positive scalar constant; then $f(\mathbb{A})$ is maximized in the space of all positive definite matrices when $\mathbb{A} = n\,\mathbb{I}_p$.*

Result 7.3.3. *Let the $\mathbf{X}_i$'s be as defined in Result 7.2.1; then the $\mathcal{MLE}$ of $\mathbb{V}$ is $\widehat{\mathbb{V}} = \mathbb{S}/n$, where*

$$\mathbb{S} = \sum_{i=1}^{n} (\mathbf{X}_i - \overline{\mathbf{X}})(\mathbf{X}_i - \overline{\mathbf{X}})'.$$

Result 7.3.4. *On substituting their maximum likelihood estimates to $\boldsymbol{\mu}$ and $\mathbb{V}$ in the likelihood function, one has*

$$\mathcal{L}(\hat{\boldsymbol{\mu}}, \widehat{\mathbb{V}}) = (2\pi)^{-np/2} |\mathbb{S}/n|^{-n/2} e^{-np/2}. \qquad (7.3.2)$$

It should be noted that $\mathbb{S} \succ \mathbb{O}$ whenever $n > p$.

7.3.2. The distribution of the $\mathcal{MLE}$s

Result 7.3.5. *Let* $\mathbf{X}_i \overset{ind}{\sim} \mathcal{N}_p(\boldsymbol{\mu}, \mathbb{V})$, $i = 1, 2, \ldots, n$; *then*

(i) $\sqrt{n}\,\overline{\boldsymbol{X}} \sim \mathcal{N}_p(\sqrt{n}\boldsymbol{\mu}, \mathbb{V})$;

(ii) $\mathcal{S} = \sum_{i=1}^{n}(\boldsymbol{X}_i - \overline{\boldsymbol{X}})(\boldsymbol{X}_i - \overline{\boldsymbol{X}})'$ *is distributed as* $\sum_{i=1}^{n-1} \mathbf{Z}_i \mathbf{Z}_i'$ *where* $\mathbf{Z}_i \overset{iid}{\sim} \mathcal{N}_p(\mathbf{0}, \mathbb{V})$, $i = 1, 2, \ldots, n - 1$, *that is,* $\mathcal{S}$ *has a Wishart distribution;*

(iii) $\mathcal{S}/(n - 1)$ *is an unbiased estimator of* $\mathbb{V}$;

(iv) $\overline{\mathbf{X}} = \sum_{i=1}^{n} \mathbf{X}_i/n$ *and* $\mathcal{S}$ *are independently distributed.*

7.3.3. The Wishart distribution

Definition 7.3.1. *The Wishart distribution.* Let $\boldsymbol{X}_i \overset{ind}{\sim} \mathcal{N}_p(\mathbf{0}, \mathbb{V})$, $\mathbb{V} \succ \mathbb{O}$, $i = 1, 2, \ldots, n$, and $n > p$. The density function of the $p \times p$ Wishart matrix random variable $\mathcal{A} = \mathcal{X}\mathcal{X}'$, where

$$\mathcal{X} = \begin{bmatrix} x_{11} & \cdots & x_{1n} \\ \vdots & \ddots & \vdots \\ x_{p1} & \cdots & x_{pn} \end{bmatrix} = [\boldsymbol{X}_1, \ldots, \boldsymbol{X}_n], \qquad (7.3.3)$$

is

$$\frac{|\mathbb{A}|^{n/2 - (p+1)/2} e^{-\frac{1}{2}\mathrm{tr}(\mathbb{V}^{-1}\mathbb{A})}}{2^{np/2}\Gamma_p(n/2)|\mathbb{V}|^{n/2}} \quad \text{for } \mathbb{A} \succ \mathbb{O} \qquad (7.3.4)$$

and 0, otherwise, where the real matrix-variate gamma is given by

$$\Gamma_p(\alpha) = \pi^{p(p-1)/4} \prod_{j=0}^{p-1} \Gamma(\alpha - j/2), \qquad (7.3.5)$$

and we write $\mathcal{A} \sim \mathscr{W}_p(n, \mathbb{V})$.

The following equality will be utilized in the derivation of Result 7.3.7.

Result 7.3.6. *Let* $\mathbb{U}_{p \times n} = (\mathbf{U}_1, \ldots, \mathbf{U}_n)$. *Then*

$$\mathbb{U}\mathbb{U}' = \sum_{i=1}^{n} \mathbf{U}_i \mathbf{U}_i'.$$

Result 7.3.7. *Let* $\mathbb{Q}_{m \times p}$ *be a matrix of rank* m *with* $m \leq p$ *(that is,* $\mathbb{Q}$ *is a full rank matrix) and* $\mathcal{Y}_{m \times n} = \mathbb{Q}_{m \times p} \mathcal{X}_{p \times n}$ *where*

$$\mathcal{X}_{p \times n} = (\mathbf{X}_1, \mathbf{X}_2, \ldots, \mathbf{X}_n)$$

and $\mathbf{X}_i \overset{ind}{\sim} \mathcal{N}_p(\mathbf{0}, \mathbb{V})$; *then* $\mathcal{Y}\mathcal{Y}' \sim \mathscr{W}_m(n, \mathbb{Q}\mathbb{V}\mathbb{Q}')$. *Thus, whenever* $\mathcal{A} \sim \mathscr{W}_p(n, \mathbb{V})$,

$$\mathbb{Q}\mathcal{A}\mathbb{Q}' \sim \mathscr{W}_m(n, \mathbb{Q}\mathbb{V}\mathbb{Q}'). \tag{7.3.6}$$

Result 7.3.8. *The moment-generating function of* $\mathcal{A} \sim \mathscr{W}_p(n, \mathbb{V})$ *is*

$$\mathcal{M}_{\mathcal{A}}(\mathbb{T}) = \mathscr{E}(e^{\mathrm{tr}(\mathbb{T}\mathcal{A})}) = |\mathbb{I} - 2\mathbb{T}\mathbb{V}|^{-\frac{n}{2}}, \tag{7.3.7}$$

where

$$\mathbb{T} = \begin{bmatrix} t_{11} & \frac{1}{2}t_{21} & \cdots & \frac{1}{2}t_{p1} \\ \frac{1}{2}t_{21} & t_{22} & \cdots & \frac{1}{2}t_{p2} \\ \vdots & \vdots & \ddots & \vdots \\ \frac{1}{2}t_{p1} & \frac{1}{2}t_{p2} & \cdots & t_{pp} \end{bmatrix}, \quad \mathbb{I} - 2\mathbb{T}\mathbb{V} \succ \mathbb{O}. \tag{7.3.8}$$

Result 7.3.9. *Reproductive Property. Let* $\mathcal{S}_1 \sim \mathscr{W}_p(m, \mathbb{V})$, $\mathcal{S}_2 \sim \mathscr{W}_p(n, \mathbb{V})$ *and* $\mathcal{S}_1$ *and* $\mathcal{S}_2$ *be independently distributed. Then* $\mathcal{S} = \mathcal{S}_1 + \mathcal{S}_2 \sim \mathscr{W}_p(m + n, \mathbb{V})$.

Exercise 7.3.

7.3.1. In connection with Result 7.3.1, show that the likelihood function of the sample has the representation given in Eq. (7.3.1).

7.3.2. Prove Result 7.3.2. First, express the determinant and trace of the matrix $\mathbb{A}$ in terms of its eigenvalues.

7.3.3. Prove Result 7.3.4.

7.3.4. Show that part (i) of Result 7.3.5 holds.

7.3.5. Show that part (iii) of Result 7.3.5 holds, i.e., the expected value of $\mathcal{S}$ is equal to $(n-1)\mathbb{V}$.

7.3.6. Prove Result 7.3.6.

7.3.7. Making use of Result 7.3.6, prove Result 7.3.7.

7.3.8. Show that Result 7.3.9 holds by making use of the moment-generating functions of $\mathcal{S}_1$ and $\mathcal{S}_2$.

7.4. Certain Test Statistics

7.4.1. Introduction

Definition 7.4.1. *Likelihood ratio statistic.* Let $\boldsymbol{X}_i = (x_{1i}, x_{2i}, \ldots, x_{pi})'$, $i = 1, 2, \ldots, n$, constitute a simple random sample drawn from a distribution whose associated parameter vector is $\boldsymbol{\theta}$, and denote the likelihood function (joint density function of the $\boldsymbol{X}_i$'s) by $\mathcal{L}(\cdot)$. Now, letting Ω and ω respectively represent the entire parametric space and a subset thereof that is specified by a null hypothesis $\mathcal{H}_0$, one can assess the validity of this null hypothesis by making use of the likelihood ratio statistic,

$$\lambda = \frac{\mathcal{M}ax_\omega \, \mathcal{L}(\cdot)}{\mathcal{M}ax_\Omega \, \mathcal{L}(\cdot)} = \frac{\mathcal{L}(\hat{\boldsymbol{\theta}}_\omega)}{\mathcal{L}(\hat{\boldsymbol{\theta}}_\Omega)},$$

where $\hat{\boldsymbol{\theta}}_\omega$ and $\hat{\boldsymbol{\theta}}_\Omega$, respectively, denote the $\mathcal{MLE}$'s of $\boldsymbol{\theta}$ in ω and Ω.

Definition 7.4.2. *Critical region.* The interval $0 < \lambda \leq \lambda_0$ is called the *critical region* at significance level α where λ_0 is such that $\int_0^{\lambda_0} p_0(\lambda)d\lambda = \alpha$, $p_0(\lambda)$ denoting the $\mathcal{PDF}$ of λ under $\mathcal{H}_0$. Thus, under the null hypothesis, $\mathcal{P}r(0 < \lambda \leq \lambda_0) = \alpha$.

Result 7.4.1. *Asymptotic Distribution of* $-2\ln\Lambda$. *Let λ be the likelihood ratio statistic obtained on the basis of the sample size n. Then, the asymptotic distribution of $-2\ln\Lambda$ is χ^2_{a-b} where a is the number of free parameters in Ω and b is the number of free parameters in ω (specified by $\mathcal{H}_0$). Free parameters are distinct parameters that have to be estimated.*

7.4.2. A test of independence

Let $\boldsymbol{X}_1, \ldots, \boldsymbol{X}_n$ constitute a simple random sample from an $\mathcal{N}_p(\boldsymbol{\mu}, \mathbb{V})$ distribution where $\boldsymbol{X}_j = (x_{1j}, \ldots, x_{pj})'$. Under the null hypothesis, it is assumed that the p components of the multivariate normal distribution are independently distributed. It can be shown that the likelihood ratio statistic for this test is

$$\lambda = \frac{\left|\frac{\mathbb{S}}{n}\right|^{n/2}}{\prod_{i=1}^{p}(s_{ii}/n)^{n/2}},$$

where

$$s_{ii} = \sum_{j=1}^{n}(x_{ij} - \overline{x}_i)^2 \quad \text{with } \overline{x}_i = \sum_{j=1}^{n} x_{ij}/n, \ i = 1, \ldots, p,$$

and

$$\mathbb{S} = \sum_{j=1}^{n}(\boldsymbol{X}_j - \overline{\boldsymbol{X}})(\boldsymbol{X}_j - \overline{\boldsymbol{X}})' \quad \text{with } \overline{\boldsymbol{X}} = \sum_{j=1}^{n} \boldsymbol{X}_j/n.$$

7.4.3. The sphericity test

Let $\boldsymbol{X}_1, \boldsymbol{X}_2, \ldots, \boldsymbol{X}_n$ be a simple random sample from an $\mathcal{N}_p(\boldsymbol{\mu}, \mathbb{V})$ population and let

$$\mathcal{H}_0 : \mathbb{V}_0 = \sigma^2 \mathbb{I},$$

where σ^2 is a scalar, not specified, $\boldsymbol{\mu}$ being unknown. Then, on letting $\nu_1, \nu_2, \ldots, \nu_p$ denote the eigenvalues of $\mathbb{S}$, one has the following test statistic:

$$\lambda^{2/np} = \frac{|\mathbb{S}|^{1/p}}{\frac{\text{tr}(\mathbb{S})}{p}} = \frac{(\prod_{i=1}^{p} \nu_i)^{1/p}}{\sum_{i=1}^{p} \nu_i/p}$$

$$= \frac{\text{the geometric mean of the eigenvalues of } \mathbb{S}}{\text{the arithmetic mean of the eigenvalues of } \mathbb{S}},$$

where λ is the likelihood ratio statistic for this test and $\mathbb{S}$ is as defined in Section 7.4.2.

7.4.4. Testing that μ equals a given vector

Definition 7.4.3. *Hotelling's T^2 statistic.* Letting $\mathbf{X}_i \overset{ind}{\sim} \mathcal{N}_p(\boldsymbol{\mu}, \mathbb{V})$, $i = 1, \ldots, n$, where $\boldsymbol{\mu}$ and $\mathbb{V}$ are unknown,

$$\overline{\boldsymbol{X}} = \sum_{i=1}^{n} \boldsymbol{X}_i / n \quad \text{and} \quad \mathcal{S} = \sum_{i=1}^{n} (\boldsymbol{X}_i - \overline{\boldsymbol{X}})(\boldsymbol{X}_i - \overline{\boldsymbol{X}})', \qquad (7.4.1)$$

Hotelling's T^2 statistic on $n - 1$ degrees of freedom is defined as follows:

$$T^2 = n(n - 1)(\overline{\boldsymbol{X}} - \boldsymbol{\mu}_0)' \mathcal{S}^{-1}(\overline{\boldsymbol{X}} - \boldsymbol{\mu}_0). \qquad (7.4.2)$$

This statistic is utilized for testing the null hypothesis, $\mathcal{H}_0 : \boldsymbol{\mu} = \boldsymbol{\mu}_0$, when $\mathbb{V}$ is unknown. It can be shown that when $\boldsymbol{\mu} = \boldsymbol{\mu}_0$, $\frac{T^2}{n-1}\left(\frac{n-p}{p}\right) \sim \mathcal{F}_{p,\,n-p}$, which denotes an $\mathcal{F}$ distribution on p and $n - p$ degrees of freedom.

Result 7.4.2. *The test statistic in terms of the likelihood ratio statistic. Letting $\mathbf{X}_1, \ldots, \mathbf{X}_n$ constitute an simple random sample of size $n > p$ from an $\mathcal{N}_p(\boldsymbol{\mu}, \mathbb{V})$ population and λ denote the likelihood ratio statistic for assessing whether $\mathcal{H}_0 : \boldsymbol{\mu} = \boldsymbol{\mu}_0$ holds when $\mathbb{V}$ is unknown, one can test $\mathcal{H}_0$ by making use of the following decreasing function of λ:*

$$T^2 \equiv (n-1)\left(\frac{1}{\lambda^{\frac{2}{n}}} - 1\right) = (n-1)\,n(\bar{\mathbf{X}} - \boldsymbol{\mu}_0)'\,\mathbb{S}^{-1}(\bar{\mathbf{X}} - \boldsymbol{\mu}_0),$$

which follows Hotelling's T^2 distribution, $\mathbb{S}$ being as defined in Section 7.4.2. Thus, large values of T^2 will lead to a rejection of the null hypothesis.

7.4.5. Multivariate analysis of variance

Consider g independently distributed p-variate normal populations having the same covariance matrix $\mathbb{V} \succ \mathbb{O}$, but possibly different mean value vectors $\boldsymbol{\mu}_1, \boldsymbol{\mu}_2, \ldots, \boldsymbol{\mu}_g$. More specifically, let $\boldsymbol{X}_{ij}$, $j = 1, \ldots, n_i$, constitute a simple random sample from an $\mathcal{N}_p(\boldsymbol{\mu}_i, \mathbb{V})$ population, $i = 1, \ldots, g$. Suppose that we wish to test whether the population mean values are equal. Accordingly, the null hypothesis is

$$\mathcal{H}_0 : \boldsymbol{\mu}_1 = \boldsymbol{\mu}_2 = \cdots = \boldsymbol{\mu}_g = \boldsymbol{\mu} \text{ (unspecified)}, \mathbb{V} \text{ being unknown.}$$

It can be shown that the likelihood ratio statistic for this test is

$$\lambda = \frac{|\sum_{i=1}^{g} \mathbb{S}_i|^{n/2}}{|\mathbb{S}|^{n/2}},$$

where

$$\mathbb{S}_i = \sum_{j=1}^{n_i} (\boldsymbol{X}_{ij} - \overline{\boldsymbol{X}}_i)(\boldsymbol{X}_{ij} - \overline{\boldsymbol{X}}_i)', \quad i = 1, \ldots, g$$

and

$$\mathbb{S} = \sum_{i=1}^{g} \sum_{j=1}^{n_i} (\boldsymbol{X}_{ij} - \overline{\boldsymbol{X}})(\boldsymbol{X}_{ij} - \overline{\boldsymbol{X}})',$$

$\overline{\boldsymbol{X}}_i = \sum_{j=1}^{n_i} \boldsymbol{X}_{ij}/n_i$ denoting the sample mean of the ith population and $\overline{\boldsymbol{X}} = \sum_{i=1}^{g} \sum_{j=1}^{n_i} \boldsymbol{X}_{ij}/n$, the grand sample mean.

Exercise 7.4

7.4.1. Explain why a likelihood ratio statistic is always between 0 and 1, $0 < \lambda \leq 1$.

7.4.2. What is initially the denominator of the likelihood ratio statistic for the test of independence described in Section 7.4.2 in terms of the likelihood function.

7.4.3. Express the numerator of the likelihood ratio statistic for the test of independence described in Section 7.4.2 in terms of a double product involving the x_{ij}'s, $i = 1, \ldots, p$; $j = 1, \ldots, n$.

7.4.4. Replacing $\boldsymbol{\mu}$ by its maximum likelihood estimator in Eq. (7.3.1) and letting $\mathbb{V} = \sigma^2 \mathbb{I}$, which corresponds to the null hypothesis for the sphericity test described in Section 7.4.3, differentiate the resulting likelihood function with respect to σ^2 in order to obtain the maximum likelihood estimate of this parameter.

7.4.5. What is the asymptotic distribution of $-2\ln\Lambda$ where Λ is the likelihood ratio statistic for testing the independence of the components of a multivariate normal vector as defined in Section 7.4.2.

7.4.6. What is the asymptotic distribution of $-2\ln\Lambda$ where Λ is likelihood ratio statistic for the sphericity test as defined in Section 7.4.3.

7.4.7. In connection with the test described in Section 7.4.4, let c denote a critical value such that one would reject $\mathcal{H}_0$ whenever $T^2 \geq c$ at significance level α. Express c in terms of a constant times a certain percentile of an $\mathcal{F}$ distribution.

7.4.8. Determine the maximum of the likelihood function under the null hypothesis specified in Section 7.4.5.

Acknowledgments

The author would like to acknowledge with thanks the funding provided by the Centre for Mathematical and Statistical Sciences, Peechi Campus, for his participation in the 2015 SERB School.

Bibliography

[1] T.W. Anderson, *An Introduction to Multivariate Statistical Analysis*, 3rd edition, New York, Wiley, 2004.

[2] A.T. Craig, Note on the independence of certain quadratic forms, *Ann. Math. Statis.* **14** (1943), 195–197.

[3] M.F. Driscoll and W.R. Gundberg Jr., A history of the development of Craig's theorem, *Amer. Statistician* **40** (1986), 65–70.

[4] H. Hotelling, Note on a matric theorem of A. T. Craig, *Ann. Math. Statist.* **15** (1944), 427–429.

[5] R. Johnson and D. Wishern, *Applied Multivariate Statistical Analysis*, 6th edition, Pearson/Prentice-Hall, Upper Saddle River, 2007.

[6] A.M. Mathai and S.B. Provost, *Quadratic Forms in Random Variables: Theory and Applications*, Marcel Dekker, New York, 1992.

[7] S.R. Searle, *Linear Models*, New York, Wiley, 1971.

Author Index

Subject Index

Printed in the United States
By Bookmasters